THE WONDERS OF MAGIC SQUARES

THE WONDERS OF
MAGIC SQUARES

BY

JIM MORAN

WITH A FOREWORD BY MARTIN GARDNER

VINTAGE BOOKS

A DIVISION OF RANDOM HOUSE · NEW YORK

A Vintage Original, January 1982
First Edition
Copyright © 1981 by Jim Moran
All rights reserved under International and Pan-American
Copyright Conventions. Published in the United States by
Random House, Inc., New York, and simultaneously in Canada
by Random House of Canada Limited, Toronto.

Library of Congress Cataloging in Publication Data
Moran, Jim.
The wonders of magic squares.
1. Magic squares.   I. Title.
QA165.M59   793.7′4   81-40093
ISBN 0-394-74798-4   AACR2

Manufactured in the United States of America

9 8 7 6 5 4 3 2

TO LINDA,
MY WIFE, WHO HELPED MORE
THAN I CARE TO ADMIT

Contents

FOREWORD

An amateur, G. K. Chesterton liked to say, frequently loves a hobby ("amateur" is French for lover) with more fire and intensity than a professional. "It is we bunglers who adore the occupation in the abstract," Chesterton wrote in an essay on croquet. "It is we to whom it is an art for art's sake." There are, he said elsewhere, "some things which a fifth-rate painter knows which a first-rate art critic does not know."

Big, blue-eyed, soft-voiced Jim Moran is an amateur recreational mathematician. And with all the love and passion of the hobbyist he has discovered some things about magic squares that even most mathematicians do not know.

Jim was a legend long before I had the pleasure of knowing him. I had, of course, read about his inspired hoaxes, his publicity pranks, and his harmless but funny practical jokes. He sold an icebox to an Eskimo, found a needle in a haystack, sat on an ostrich egg until it hatched, changed horses in the middle of a stream in Nevada, tested the effects of booze on a hoot owl, reenacted the Battle of Bunker Hill, proved that a bull in a china shop won't break a thing, and acquired a large collection of recordings of different kinds of toilet flushes.

Disguised as an Arabian crown prince, Moran once "accidentally" spilled a bagful of synthetic gems on a crowded nightclub dance floor. To publicize a Broadway play he rode through Manhattan in the back seat of a London taxi with an orangutan in the driver's seat. He showed that there is no difference between Florida and California sunshine by allowing half his body to tan in one state, half in the other. He called on Gelett Burgess (who wrote "I never saw a purple cow . . .") leading a cow that had been dyed purple. I could go on and on. But if a psychic had told me I would someday meet the Magnificent Moran through a common interest in magic squares, I would have thought the prediction as improbable as one of Moran's own escapades.

And yet there had been hints of Jim's interest in mathematics. On television I had seen him demonstrate his rotating Ames window, which through tricks of projective geometry seems to pass through a solid steel bar. On Jack Paar's show I had watched Jim and Jack go through hilarious acrobatics while Jim tried to remember, or pretended to try, how to solve an old topological puzzle involving two people linked together by strings tied to their wrists. On a radio show I had heard Jim sing the great lyrics he had written for the tune of "Sobre las olas" (Over the Waves)—lyrics that paid tribute to that monstrous mathematical structure called the George Washington Bridge.

When we first met, Jim was still in the euphoric flush of having just discovered magic squares, and they had aroused in him an intense desire to learn some basic number theory. What, he wanted to know, were those mysterious exclamation marks he kept seeing in printed equations? They are, I said, symbols for factorials. "What the hell are factorials?" Jim took notes while I explained. A few months later he was discovering, all by himself, ways to construct large magic squares that had been the secret techniques of early mathematicians of the highest ranks.

No one but a novice with a deep sense of wonder and curiosity about numbers, and a wild sense of humor, could have written this entertaining book. It is the only book I know that manages not only to explain magic squares to mathematical illiterates but also to hook their interest in these fascinating curiosities. No other book conveys in such simple and clear language, with so much zest and wit, the eerie beauty of magic squares, the striking visual patterns concealed inside them, and the sheer fun of making them and searching out their incredible properties. And who but Jim Moran could have imagined connections between magic squares, tic-tac-toe, genetics, atom bombs, and eggs fu yung?

—MARTIN GARDNER

THE WONDERS OF MAGIC SQUARES

PART ONE

INTRODUCTION

The construction of magic squares is a mind-blowing diversion of great antiquity. Their mystical significance was known in China and India centuries before the time of Christ. They were engraved on pendants of stone, metal, or wood and worn on necklaces as talismans or amulets to bring good fortune or to ward off malign spirits or the evil eye.

They are still being excavated by archeologists and are still to be seen among the ruins of ancient oriental cities, carved on doors of crumbling houses or above the entrances to forgotten temples.

Those few ancients who knew the carefully guarded secrets of their construction were held in awe, fear, and reverence by ordinary superstitious people. The holders of the secrets were understandably not averse to using their knowledge for personal advancement. Many used this means to gain power and influence in government and religion.

The earliest discussion of a magic square in the West occurs in the works of Theon of Smyrna about 130 A.D. Arabic astrologers in the ninth century were using magic squares in their calculations of horoscopes, but it was largely through the writings of Emanuel Moschopulus, a Greek mathematician and grammarian who lived in Constantinople about 1300 A.D., that the knowledge of magic squares eventually spread throughout the western world and fascinated the great minds of the Renaissance. His work in manuscript is in the National Library in Paris.

Cornelius Agrippa (1486–1535), a German physician and theologian, constructed various magic squares that were associated with the then-known seven astrological "planets"—Saturn, Jupiter, Mars, the sun, Venus, Mercury, and the moon.

One of the world's most famous engravings, *Melencolia* by Albrecht Dürer, depicts a 4 × 4 magic square. The date of the work (1514) is indicated in the two middle boxes of the bottom row.

Benjamin Franklin, one of the outstanding magic-square freaks of his time, has a chapter devoted to him in this book.

Pythagoras in the fifth century said that numbers govern the origin of all things and that the law of numbers is the key that unlocks the secrets of the universe.

Magic squares brilliantly reveal the intrinsic harmony and symmetry of numbers; with their curious and mystic charm they appear to betray some hidden intelligence that governs the cosmic order that dominates all existence. They have been compared to a mirror reflecting the symmetry of the universe, the harmonies of nature, the divine norm. It is not surprising that they have always exercised a great influence on thinking people.

How i got hooked on magic squares

I got hooked on magic squares one night about five years ago when a visiting friend drew a diagram like this:

and asked me if I knew anything about magic squares.

"No," I said, "I never heard of 'em."

"Well, here's a little mathematical problem that might interest you," he said. "Take the numbers from one through nine and arrange them in the nine boxes in this diagram so that the three numbers in each row, each column, and the corner-to-corner diagonals will add up to the same total. This is called a magic square."

"I'm no good with numbers," I said, "but at least I can add. This looks easy."

But as it turned out, it wasn't that easy.

In the first place, I'd forgotten all the math I had studied in high school. I did remember that algebraic equations when plotted on graph paper made pretty, wavy, symmetrical lines. This impressed me very much at the time, but I wouldn't know how to do it today. I know that $8 + 6 = 14$, but I'm never sure unless I count on my fingers. Balancing my checkbook once a month takes me half a day. One look at an income tax form is enough to drive me up the wall. So I'm not exactly what one might call a mathematician.

Despite my shortcomings, I went to work trying to make the 3×3 magic square.

To appreciate fully the wonder of magic squares, I suggest that the reader stop here and, starting from scratch like I did, try to make the 3×3 magic square.

I first tried to find the common total that each row and column and the diagonals should add up to. After an hour of juggling the numbers, I still hadn't even figured out this first step. It was getting late, my friend was getting bored, so

I gave up and asked him to show me the secret. He did and I was surprised at the simplicity of the method once I got the hang of it.

He told me that this method could be used to make "odd-order" magic squares of any size. "Odd-order," he explained, meant 3 × 3, 5 × 5, 7 × 7, 9 × 9, etc.

After my friend left, I couldn't wait to try out the system on 5 × 5, 7 × 7, and 9 × 9 squares. Constructing the new magic squares kept me up most of the night.

I felt strangely pleased and exhilarated after completing these first magic squares. The feeling is difficult to describe, but I think it had something to do with dissonance and harmony. When I succeeded in arranging the numbers in magic-square order, a dissonance seemed to be replaced by pure harmony. Musically speaking, I felt that I had found the "lost chord." That night I slept with a deep feeling of satisfaction and accomplishment.

The next night I looked at the magic squares I had made. I was still satisfied, but not entirely. By using the system my friend had shown me, I could make only *one arrangement* of the numbers for each of the odd-order squares. A nagging hunch kept telling me that there *must* be other methods whereby I could use a *different* arrangement of the numbers to achieve the desired result—and what about the "even-order" squares, 4 × 4, 6 × 6, 8 × 8?

I tried to make a 4 × 4, but the method I had been using wouldn't work. I was determined to make a 4 × 4 magic square—after all, there are only sixteen numbers to juggle—but after three weeks of juggling, I still couldn't make one. (Before continuing, I suggest the reader try to make a 4 × 4 magic square.)

This was the beginning of my addiction!

I then consulted the major encyclopedias; they helped some but not much. They showed a 4 × 4 magic square but didn't explain how it was made. They gave me a brief history of magic squares going back to ancient China. They cited famous people who had been hooked and showed a 16 × 16 magic square constructed by Benjamin Franklin, who claimed this to be "the most magically magical of any magic square ever made by any magician." They presented a few methods for constructing both odd- and even-order squares, but their explanation of the construction methods was given in mathematical language and symbols (all Greek to me!).

I finally made my first 4 × 4 all by myself and gave a party to celebrate the event.

My habit grew. I started discovering more and more curious and unexpected things having to do with square arrangements of numbers. I didn't know it at the time but I was following the footsteps of men who died before the time of Christ. Each new discovery brought more questions. Magic squares started to interfere with my ways of making a living. I became a recluse and stopped entertaining. I turned down dinner and party invitations and gave up going to shows and movies—all because magic squares were *more interesting*. I haunted the New York Public Library and devoured every book that had any mention of magic squares. (I devoured them but I'm afraid my digestion wasn't too good.) I studied some math and picked up a few helpful clues.

The books displayed hundreds of examples of completed magic squares of all sizes. I read that there are 880 ways to arrange the 16 numbers to make 4 × 4 magic squares (it took me three weeks to find *one*). A recent computer analysis indicates that there are over *275 million* different 5 × 5 magic squares. But most of the books I found in the library were written long ago, and all seemed to have been written by mathematicians for mathematicians. They told me little that I could understand about the many methods of construction.

For over a year, my consuming curiosity led me to discover more and more fascinating aspects of magic squares. During all this time, it had never occurred to me that I might someday write a book on the subject.

But as I got deeper and deeper into the subject, it slowly dawned on me that even though magic squares have been of great interest to countless thinking people for some three thousand years, nobody, to my knowledge, had yet written a simple book on the subject that could be understood by anyone who is able to add up a small column of figures. At this point, a light bulb went on in my head and I decided to attempt to write such a book. This is it.

P.S. Again, I urge the reader to attempt to make the 3 × 3 and 4 × 4 magic squares before continuing further.

A 5 X 5 MAGIC SQUARE

17	24	1	8	15
23	5	7	14	16
4	6	13	20	22
10	12	19	21	3
11	18	25	2	9

The above is a simply constructed magic square. It is called magic because the five numbers in each row, each column, and the two corner-to-corner diagonals add up to the same total—65. In this square, there are *millions* of ways to arrange the numbers 1 through 25 to produce the same result, yet an average intelligent individual without a knowledge of the systems involved might work for days and still be unable to discover even one solution.

Use blank graph paper to make magic squares

CAN YOU DO THIS?

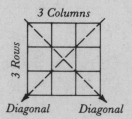

Empty Diagram for 3 × 3 Magic Square (9 Boxes)

If you can arrange the numbers from 1 through 9 in the nine boxes above so that the three numbers in each row, each column, and each corner-to-corner diagonal will add up to the same total, you are well on your way to becoming a magic-square freak.

There are eight different arrangements that will be shown later, but if you really want to enjoy this trip—and it really *is* a trip—hang in there and don't turn to the next page until you have done your best.

Use blank graph paper opposite.

If you have not yet arrived at a solution, here is the first clue:

The total of the three numbers in each column, row, and corner-to-corner diagonals in a 3 × 3 magic square is 15. This total is called *the constant* for that square.

Three ways to find the constant of magic squares of any size:

A. Add the first and last numbers of the sequence and multiply this total by the number of boxes in a row. Divide this figure by 2 to get the constant.

Examples: The *3 × 3 square* contains nine boxes and the sequence 1 through 9 is used to fill the boxes. Add the first and last numbers of the sequence: 1 + 9 = 10. Multiply 10 by the number of boxes in a row (3): 3 × 10 = 30. Divide 30 by 2 to get 15, the constant of the 3 × 3 magic square.

4 × 4 square: 1 + 16 = 17; 17 × 4 = 68; 68 ÷ 2 = 34, the constant.

5 × 5 square: 1 + 25 = 26; 26 × 5 = 130; 130 ÷ 2 = 65, the constant.

This is the quickest and best method of finding a constant, but a knowledge of other methods (see below) still proves valuable in understanding the beauty of magic squares.

B. Add up all of the numbers in the sequence and divide this total by the number of boxes in a row.

Examples: 3 × 3 magic square—sequence 1 through 9. Add: 1 + 2 + 3 + 4 + 5 + 6 + 7 + 8 + 9 = 45. Divide 45 by 3 (the number of boxes in a row) to get the constant, 15.

4 × 4—sequence 1 through 16. Numbers of sequence added is 136. 136 ÷ 4 = 34 (constant).

5 × 5—sequence 1 through 25, total 325. 325 ÷ 5 = 65 (constant).

Etc.

C. Make a diagram of any size magic square. Place the number 1 in the box at the upper left corner. Fill in the remaining boxes of the top row with the numbers in their natural sequence (1, 2, 3, etc.). Continue with second row in the same manner until the last number is placed in the box at the lower right corner. Add the numbers in *either* corner-to-corner diagonal. The total will be the constant for that square.

Examples:

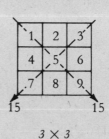

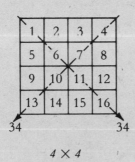

3 × 3   4 × 4   5 × 5

The occult aspects of numbers begin to reveal themselves even in the most simple natural-sequence arrangement of numbers as in the squares above. Does it not seem strange that the sum of the numbers in *both* diagonals of each square is identical? More careful observation will disclose the peculiar fact that the sum of any two numbers equidistant from the center is the same as the sum of the first and last numbers of the series used in that square. Example: The sum of the first and last numbers in the 5 × 5 square is 26. 9 and 17 are equidistant from the center. When added, they total 26. 26 is also the sum of any other pair of numbers equidistant from the center, such as 10 and 16, 2 and 24, 5 and 21, 6 and 20, 3 and 23, etc.

If you still haven't been able to construct the 3 × 3 magic square, here is the last clue: The number 5 *must* be placed in the center box. (5 is the middle number in the series 1 through 9.)

	5	

Keep trying. The next chapter will explain how to make the 3 × 3 magic square, but try on your own before you look.

THE CONSTRUCTION OF ODD-ORDER MAGIC SQUARES

Method A:
THE STAIRCASE METHOD

This well-known method invented by La Loubère may be used to construct odd-order magic squares of any size—3 × 3, 5 × 5, 7 × 7, etc.

Instructions:

1. Construct a blank diagram (see below).

2. Imagine the above diagram surrounded by identical diagrams (see below—dotted lines indicate imaginary diagrams).

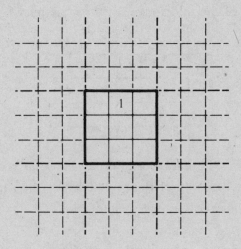

3. *Rule one:* Always start by placing the number 1 in the middle box of the top row of the magic square under construction (see above).
4. *Rule two:* Place each succeeding number in the box diagonally upward to the right—unless this box is occupied by a previously placed number.
5. Since the number 1 occupies the middle box of the top row, we must now, if we follow rule two, place the number 2 diagonally upward to the right, which places it in the lower right corner box of the imaginary 3 × 3 square directly above the square under construction.

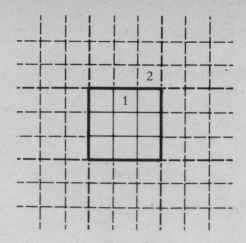

6. Since we have placed the number two *outside* of the magic square under construction, we must find the proper place for it *inside*.

7. Note that 2 is temporarily occupying the lower right corner box of the imaginary square above. We must move it to the same relative position (lower right corner box) inside the magic square under construction (see below).

8. Following rule two, the number 3 is temporarily placed diagonally upward to the right of 2. This again lands our number *outside* of the square under construction and into the first box of the middle row of the imaginary square (see below).

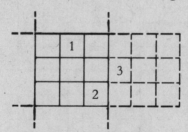

9. To place the number 3 where it belongs *inside* the square under construction, we must move it into the first box of the middle row—the same relative position it occupies in the imaginary square (see below).

	1	
3		
		2

10. It is obvious that we cannot follow rule two and place the number 4 in the box diagonally upward to the right because that box is already occupied by the previously placed number 1. This brings us to the final rule.

11. *Rule three:* When the box diagonally upward to the right is blocked by a previously placed number, place the succeeding number in the box directly below the last number placed; 4 is now placed below 3 (see below).

	1	
3		
4		2

12. Following rule two, the numbers 5 and 6 are placed in the boxes diagonally upward to the right from the number 4 (see below).

	1	6
3	5	
4		2

13. Following rule two, the number 7 must be placed in the box diagonally upward to the right, which lands it in the lower left corner box of the imaginary square diagonally upward to the right of the square under construction.

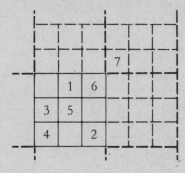

14. When we attempt to move 7 from its temporary location in the lower left corner box of the imaginary square to its relative position within the square under construction, we find that we are blocked by the previously placed number 4. Following rule three, we must place 7 in the box directly below 6 (see below).

	1	6
3	5	7
4		2

15. Following rule two, the number 8 must be placed diagonally upward to the right of number 7, which lands it in the upper left corner box of the imaginary square to the right of the square under construction (see below).

	1	6	8
3	5	7	
4		2	

16. The number 8 must be moved from its temporary position in the imaginary square to the same relative position (upper left corner) in the square under construction (see below).

8	1	6
3	5	7
4		2

17. The number 9, when placed diagonally upward to the right of 8, will occupy the middle box of the lower row of the imaginary square above. By placing 9 in the same relative position in the square under construction, we have completed the 3 × 3 magic square (see below).

8	1	6
3	5	7
4	9	2

Note: The numbers in each row, column, and corner-to-corner diagonals of the 3 × 3 square total 15.

The opposite pairs of numbers equidistant from the center (8 and 2) (6 and 4) (7 and 3), etc., total 10—the sum of the highest and lowest numbers in the sequence (1 and 9). *These pairs are called complementary numbers.* Complementary pairs of numbers are shown in natural sequence order below.

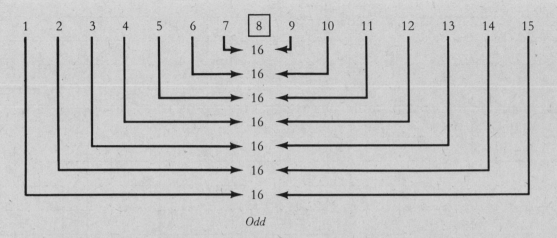

Odd

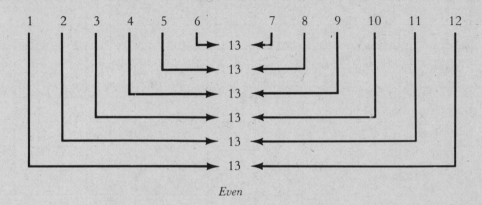

Even

HOW TO HAVE FUN
WITH THE 3 X 3

1. Make a blank 3 × 3 diagram and ask your victim to place any number between 10 and 100 in any box.
2. Let's assume he puts 68 in the upper right box.

3. You rapidly fill in the eight remaining boxes with eight numbers.

70	63	68
65	67	69
66	71	64

4. Ask your victim to add up each row, column, and corner-to-corner diagonal. He will be astounded to find that they all add up to 201. He may even assume you are some sort of math genius. If he thinks it looks easy, ask him to try it.
5. *The secret:* To do this trick, you must memorize the 3 × 3 magic square.

8	1	6
3	5	7
4	9	2

6. It will be seen that 68 has been placed in the box occupied by 6 in your memorized square—in other words, he has added 62 to the value of that box.
7. Now all you have to do is to add 62 to each of the other boxes of your memorized 3 × 3 and you will have made a magic square using the series 63 through 71, which has a constant of 201.

CONDENSED RULES FOR METHOD A (THE STAIRCASE METHOD) FOR ODD-ORDER MAGIC SQUARE OF ANY SIZE

Rule one: Always begin by placing the number 1 in the middle box of the top row.

Rule two: Place each succeeding number in the box diagonally upward to the right, unless this box is occupied by a previously placed number.

Rule three: When the box diagonally upward to the right is occupied by a previously placed number, place the succeeding number in the box directly below the last number placed. This is known as a blocked move.

The above rules are given for convenience and should be followed only after thoroughly understanding the details of method A. See 5 × 5 example below.

17	24	1	8	15
23	5	7	14	16
4	6	13	20	22
10	12	19	21	3
11	18	25	2	9

Historical evidence places the origin of magic squares in ancient China, but exactly where, when, and by whom they were invented will probably never be known for sure. When facts about ancient times are skimpy or unavailable, historians frequently attempt to fill in the blank spaces by educated guesswork or by putting two and two together. The following story was most likely formulated in this way.

While I was visiting Hong Kong, the subject of magic squares came up in a conversation with my young friend Charlie Kwan, who after recently completing his college education in America now manages his family's export business in Kowloon. Charlie laid a tale on me which he said was told him years ago by his great-grandfather. At my request, he has kindly written the story, which is presented herewith.

"THE INVENTION OF THE MAGIC SQUARE"

BY CHARLIE KWAN

Early in the reign of the Chinese emperor Yu in the fifth century B.C., there lived in the remote province of Kan-su a remarkable youth, Wang Fu Yung, whose obsession and principal joy in life was the game of tic-tac-toe. Before reaching his sixth year, he had defeated his father, the wealthy rice merchant Yung-Lo, with such regularity that the honorable gentleman courteously declined his son's daily invitations to play further games.

When he was fourteen, Fu Yung won the Royal National Open Tic-Tac-Toe Tournament in Peking for the third straight year by defeating the most eminent scholars of the empire, including Chang-yang, the crafty minister of finance, who, suffering loss of face by his defeat, used his many wiles in attempts to discredit Fu Yung in the eyes of the emperor.

The wise and benevolent emperor saw fit to resolve the spreading discord by pronouncing Fu Yung the "all-time, all-time tic-tac-toe champion of the world," presented him with the royal jade-inlaid tic-tac-toe board, and canceled the yearly tournaments for the lifetime of the winner.

The hazards of attaining great success too early in life are truly expressed in the Chinese fortune-cookie wisdom "Man with one shot shoot too early—no good." Fu Yung found himself ensnared in this bind. As his fame spread throughout the land, fan mail, handsome gifts, social invitations, offers of jobs, requests for endorsements, marriage proposals, and so forth were delivered each day in huge vermilion baskets entwined with golden dragons. When he ventured out, his beautifully embroidered silk robes became further embellished in the style of Jackson Pollock with ink from the out-held brushes of aggressive autograph hounds and grabby groupies who mobbed him wherever he went.

Fu Yung gracefully accepted the homage of his nation with modest dignity, but deep down he was restless and unhappy because nobody wanted to play tic-tac-toe with him any more.

One beautiful spring day while in a particularly sad and troubled state of mind, Yung gathered up his writing implements and his treasured jade-inlaid tic-tac-toe board and retired to a secluded nook in his father's garden. Contemplating his woes beneath a weeping willow and surrounded by blossoming plum trees, he sought some poetic inspiration to relieve his misery. "Have I already climbed so high up the ladder of success that I have no place to go but down?" he mused. "Am I a one-shot man like the fortune cookie says, or do I still have a few more shots left in me?" As his mind drifted wearily, he looked down at his board in his lap and found that he had unconsciously drawn a square around his tic-tac-toe diagram (see below).

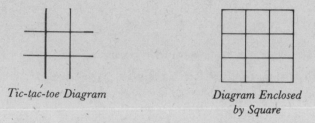

Tic-tac-toe Diagram   Diagram Enclosed
                                by Square

He concentrated on the possible hidden meanings of the square. Maybe it means I've boxed myself in, he thought. While he was pondering this, his eyes were again drawn to the board and he was surprised to find out that he had absentmindedly written the numbers from 1 through 9 in the nine boxes within the square (see sketch).

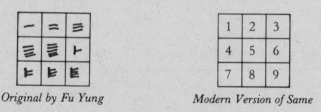

Original by Fu Yung   Modern Version of Same

This must be an omen of some kind, but what can it mean? thought Fu Yung as he looked at the board. The more he studied his simple arrangement of numbers, the more strange things he noticed. For example:

A. The middle column of three numbers and the middle row of three numbers each add up to the same total: 15.
B. The two corner-to-corner diagonals both add up to the same total: 15.
C. Each pair of opposing numbers equidistant from the center total the same: 10.

Somebody or something is trying to smuggle a message through to me, thought Fu Yung as he digested his observations. I wonder what would happen if I made a larger square with more boxes? So he drew a 5 × 5 diagram and filled the twenty-five boxes just like before.

一	二	三	亖	亖
卜	卜	卡	卡	으
上	上	业	业	业
上	上	上	上	으
土	土	业	业	业

Original by Fu Yung

1	2	3	4	5
6	7	8	9	10
11	12	13	14	15
16	17	18	19	20
21	22	23	24	25

Modern Version of Same

Sure enough, the numbers in the middle column, the middle row, and in each of the corner-to-corner diagonals all added up to the same total (but this time to 65).

Also the opposing pairs of numbers equidistant from the center box (21, 5; 3, 23; 9, 17; 6, 20; etc.) each added up to the same total: 26 (which is the sum of the first and last numbers).

Fu Yung tingled with excitement and felt sure he was on the verge of a great discovery of some kind.

For three feverish weeks without knowing why he pursued his hunch and kept shifting the nine numbers around to different positions on his board and carefully studying each new arrangement. Despair had begun to set in when the great mystery revealed itself. Fu Yung had made the world's first magic square!

At first he was too stunned to move, but when the full impact of his discovery struck him, he leaped into the air with an ear-splitting yell.

If he had been a Greek, he would have yelled "Eureka!"

If he had been an Italian, he would have yelled "Mamma mia!"

If he had been a Mexican, he would have yelled "Caramba!"

If he had been an American, he would have yelled "Wow!"

But he was Chinese, so he yelled "Yah-hoo! Yah-hoo! Yah-hoo!" like a fiend, and dashed around the garden, hugging and kissing all the trees.

His honorable old man, hearing all the commotion, laid down his abacus and glided into the garden. "I think my boy has done flipped his pigtail" was his first thought.

"What's all the ruckus about, son?"

Fu Yung bowed. "I've just invented a magic square, honorable sire," and showed his father how all the columns, rows, and diagonals added up to the same total (see sketch).

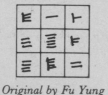

Original by Fu Yung

Modern Version

"Congratulations, son, I know how hard you have worked to produce this most interesting gizmo, but what good is it?"

"I haven't the slightest idea," replied Fu Yung, "but it turns me on."

"Good! A man with nothing to turn him on is noplace," said his father, quoting a fortune cookie. And with that bit of kindly wisdom, he shuffled back to his accounts.

Returning to his nook, Fu Yung was surprised to see a large purple turtle kneeling before the magic square. Another omen, thought Fu Yung, and he was seized with an unaccountable urge to carve the magic square on the back of the purple turtle, which he proceeded to do with the obvious consent of the turtle. When he was finished, the purple turtle gave him a lingering look of sincere gratitude and cut out for parts unknown.

Fu Yung spent the next year perfecting methods and making larger magic squares.

Meanwhile, the purple turtle, having taken a shortcut to Peking, wandered into the gardens of the Imperial Palace, where the emperor and the entire court were gathered in festive celebration of the new year.

"Hey, man!" said the first gongist of the Royal Orchestra. "Get a load of this crazy purple turtle!" The emperor and his chief advisers hurriedly gathered around and studied the magic square.

"I must know the significance of this miracle," said the emperor. The mandarins of the court went into muddled huddles. The Grand Vizier shook a fistful of marked bamboo sticks and threw them onto the ground, looked at the markings, referred to the indicated hexagram in his *I Ching* and said, "Your most illustrious highness, according to the ancient wisdom of 'The Book,' there reposes somewhere in your empire a great numbers genius."

"I must find him," said the emperor, "for whereas the royal account books kept by the nimble-fingered minister of finance are written in black ink, the diminishing stacks of gold coins in the imperial treasury leads this one to believe that red would be a more appropriate color. I suspect there's a chink in the woodpile someplace."

The emperor thereupon decreed a National Open Magic Square Tournament to be held in Peking on the first day of spring.

The meritorious Fu Yung won the contest hands down with a brilliant array of 4 × 4's, 5 × 5's, as well as a composite 9 × 9. No other contender could make even one magic square, except Chang-yang, the corrosive minister of finance, who cunningly came up with a different looking 3 × 3. Fu Yung took out his silver pocket mirror and showed the emperor that Chang-yang's 3 × 3 was merely a mirror image of the famous purple-turtle square. The devious Chang-yang, trapped in his crummy deception, resigned his lucrative office, shaved his head in humiliation, left the capital in disgrace, and began life anew as an assistant trainer of rural elephants.

The emperor celebrated Fu Yung's achievement by throwing the greatest Mardi Gras wingding in the history of the Orient. A dazzling display of ingenious fireworks lit up the night sky for two weeks. Some of them portrayed magic squares as well as the likeness of the inventor. The thundering explosions and random bursts of vivid color frightened fierce dragons as far away as the mountains of Outer Mongolia and caused terror-stricken coolies and beasts of burden in distant areas to drop their loads in unexpected places.

The climax of the festivities was a sit-down dinner for ten thousand guests. Fu Yung was seated at the right hand of the emperor, who announced the betrothal of his youngest daughter, the gorgeous and bewitching Princess Ming Chu, to the guest of honor. He then announced the appointment of Fu Yung as minister of finance and keeper of the golden keys to the royal treasury.

Last, but not least, the emperor's private chef created a fabulous omelette of two thousand hummingbird eggs for Fu Yung and the royal family. The emperor pronounced it a culinary masterpiece and named the dish Eggs Fu Yung in honor of the inventor of the world's first magic square.

DISGUISED MAGIC SQUARES

All magic squares can be turned in different directions and mirror images also can be made to produce seven additional magic squares, which at first sight appear to be different from the original but are merely the original magic square in disguise.

Generally in counting different magic squares, the disguised squares are ignored.

Disguised magic squares are difficult to detect from the originals merely by looking at the numbers. This difficulty may be overcome by drawing sequence designs that present a graphic picture, making it much easier to detect the disguised squares.

Sequence designs are generally made by drawing a straight line from 1 to 2 to 3 to 4 to 5, etc., between all the numbers in a magic square. The design is completed by drawing a final line between the last and first numbers.

A little arrow has been placed on top of the number 1 of the original 3 × 3 square below to indicate the original upright position of the square. By observing the position of the arrow on the disguised squares, we can see how the original square has been turned or mirror images made to produce the disguised squares.

It can be seen that the eight 3 × 3 sequence designs below are identical except for rotations-reflections (R.R.'s).

3 X 3 DISGUISED MAGIC SQUARES (R.R.'S)

ROTATIONS AND REFLECTIONS SHOWN BY SEQUENCE DESIGN*

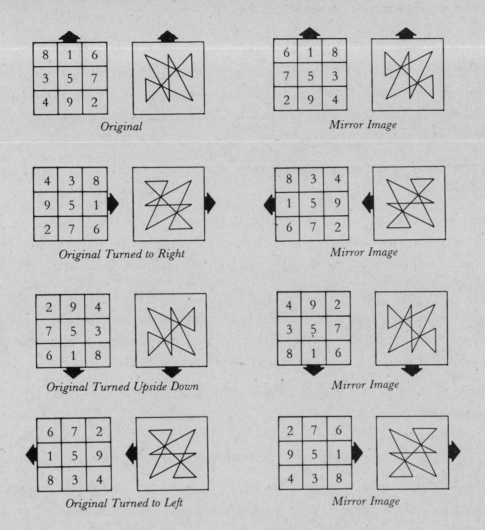

8	1	6
3	5	7
4	9	2

Original

6	1	8
7	5	3
2	9	4

Mirror Image

4	3	8
9	5	1
2	7	6

Original Turned to Right

8	3	4
1	5	9
6	7	2

Mirror Image

2	9	4
7	5	3
6	1	8

Original Turned Upside Down

4	9	2
3	5	7
8	1	6

Mirror Image

6	7	2
1	5	9
8	3	4

Original Turned to Left

2	7	6
9	5	1
4	3	8

Mirror Image

All squares above except the original are disguised squares.

*Sequence designs are made by drawing a line from 1 to 2 to 3 to 4 and so forth. The final line is drawn between the highest and the lowest numbers.

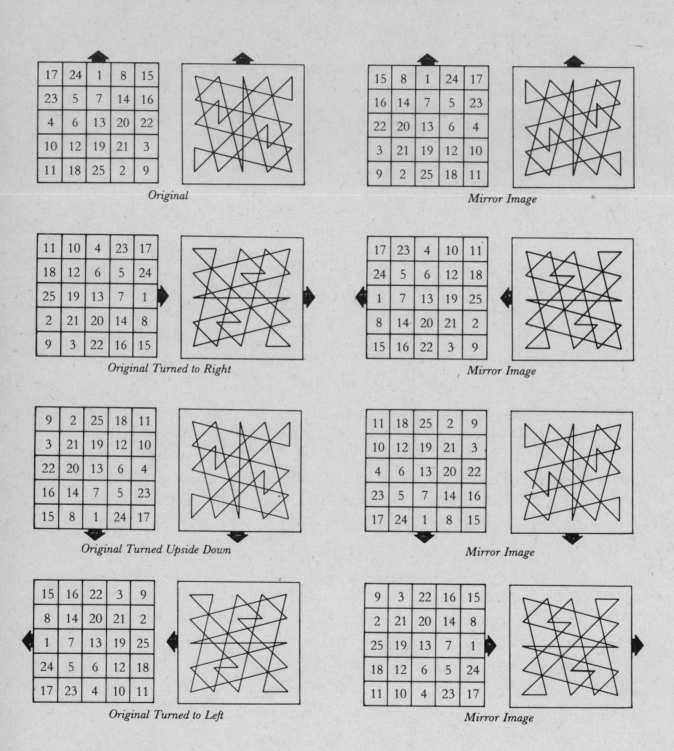

Original

Mirror Image

Original Turned to Right

Mirror Image

Original Turned Upside Down

Mirror Image

Original Turned to Left

Mirror Image

All squares above except the original are disguised squares.

⊚DD-ORDER MAGIC SQUARES

Magic squares with an odd number of boxes are usually constructed by methods that differ from those governing the construction of squares having an even number of boxes, so these two classes, known as *odd-order* and *even-order* magic squares, will be considered separately.

30	39	48	1	10	19	28
38	47	7	9	18	27	29
46	6	8	17	26	35	37
5	14	16	25	34	36	45
13	15	24	33	42	44	4
21	23	32	41	43	3	12
22	31	40	49	2	11	20

17	24	1	8	15
23	5	7	14	16
4	6	13	20	22
10	12	19	21	3
11	18	25	2	9

8	1	6
3	5	7
4	9	2

3 × 3
Constant 15

5 × 5
Constant 65

7 × 7
Constant 175

The odd-order magic squares above are all constructed by the same method and have certain things in common that are noteworthy. (A) The number 1 is in the middle box of the top row. (B) The last number of the sequence is in the middle box of the bottom row. (C) The middle number of each sequence is in the center box of the square using that sequence. (D) The sum of any pair of numbers diametrically equidistant from the center of the square is identical.

Example: The center box of the 7 × 7 square contains the number 25, which is the middle number of the sequence 1 through 49 of that square. The pairs 24 and 26, 15 and 35, 19 and 31, 47 and 3, 5 and 45, and so forth are all diametrically equidistant from the center, and when added, total 50. The number 50 is the sum of the first and last numbers (1 through 49) of the 7 × 7 magic square.

A MAGIC-SQUARE
FREAK SPEAKS

AS TOLD BY ALBERT CARLO*

My buddy, Billy, who just got hooked, didn't know I was a card-carrying magic-square freak when he fell by my pad the other night with Linda to show me how smart he is. I don't mind telling you Billy bugs me a bit. He beats me at ping-pong, Indian wrestling, bowling, shooting pool, and other stuff, and he never lets me forget it, especially when girls are around. And this night, to top it all off, he's got to bring along the yummy Linda. I met her first and we were getting along real good until he came along and started putting me down.

I knew he was trying to give me the shaft again in front of Linda when he sat down and drew a 9 × 9 diagram and then filled in the boxes as fast as he could write the numbers. "Add up any one of these columns," he says. I saw right away that he was using the simple old up-the-staircase method, but I went along with him, playing dummy and added up the middle column.

"What total you got?" he asked.

"Looks like 369," I replied.

"You're one hundred percent right," he says. "Now add up any row."

I screwed up my face, slowly added up the bottom row, and acting very surprised, said, "You don't mean to tell me that every row and every column add up to the same 369?"

"That ain't nothing," said Bill. "Add up the diagonals."

When I got finished, I played like I was flabbergasted and said, "That's super fantastic, Billy. How do you do it?"

"I can't explain it," says Billy. "I've got a head like a computer. I just seem to *know* where to put the numbers."

I begged him to let me examine the square for a minute.

"Study it all you want," he says. "But if you ain't got a computer head, forget it." Then he reaches over and squeezes Linda's hand and she gives him one of those looks I wish she'd give me again.

---

*Albert Carlo, age sixteen, the son of a good friend, heard I was interested in magic squares and asked to meet me. I was pleasantly surprised to find that he was a magic-square freak, and we spent a lively evening discussing methods. Before he left, he told me a story that I think deserves retelling. In writing it down, I have attempted to capture his mode of expression.

While I was looking it over, Billy made another 9 × 9 blank diagram, handed it to me, and took his original back. "We gotta cut," he says, "Work on that diagram and I'll be back in a few days to see how you're doin'."

"Just a minute, Billy, give me that pencil." I turned so he couldn't see what I was doing and within a minute, I had filled the eighty-one boxes with a brand-new arrangement of numbers and handed it to Billy. He glanced at it, and seeing it wasn't anything like his original, said, "Man, you've got these numbers scattered all over, but the point is to make a magic square. The columns, rows, and diagonals have to add up to 369."

"Add 'em up," I said.

While adding them up, Billy turned pale green. Then he gave Linda a panicky look and turned red. Then he narrowed his eyes and said to me, "It's impossible. How did you do it?"

"I can't explain it, Billy my boy. All I know is, when I close my eyes and concentrate, the numbers light up in the boxes like neon signs. All I have to do is open my eyes and fill 'em in. You know, since I've been on this biofeedback kick, all kinds of things are happening with my head."

When Linda said she wanted to stay and talk to me awhile, Billy cut out in a huff. I leveled with Linda and showed her the system, and now she's my old lady. Billy hasn't been back, but I heard he bought a $99.50 biofeedback machine. Here are the two squares, Billy's and mine.

47	58	69	80	1	12	23	34	45
57	68	79	9	11	22	33	44	46
67	78	8	10	21	32	43	54	56
77	7	18	20	31	42	53	55	66
6	17	19	30	41	52	63	65	76
16	27	29	40	51	62	64	75	5
26	28	39	50	61	72	74	4	15
36	38	49	60	71	73	3	14	25
37	48	59	70	81	2	13	24	35

Billy's

71	64	69	8	1	6	53	46	51
66	68	70	3	5	7	48	50	52
67	72	65	4	9	2	49	54	47
26	19	24	44	37	42	62	55	60
21	23	25	39	41	43	57	59	61
22	27	20	40	45	38	58	63	56
35	28	33	80	73	78	17	10	15
30	32	34	75	77	79	12	14	16
31	36	29	76	81	74	13	18	11

Albert's

It is obvious that Billy used method A (the staircase method) to construct his 9 × 9 magic square.

Albert, using a clever variation of the same method, constructed a 9 × 9 composite magic square, which is explained below.

ALBERT'S 9 × 9 COMPOSITE MAGIC SQUARE: VARIATION OF METHOD A

To understand properly the construction of Albert's 9 × 9 composite magic square, it must be thought of in a different way: it is not only a 9 × 9 square containing 81 boxes but also an arrangement of nine 3 × 3 squares, each of which contains nine boxes as shown below. The 9 × 9 may now be looked upon as a 3 × 3 with nine boxes. Our first move at this point is to lightly pencil in the numbers of the 3 × 3 magic square, as shown in method A (the staircase method).

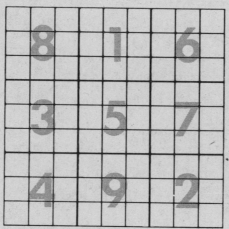

A 9 × 9 Square Divided by Heavy Lines to Look Like 3 × 3 Square
Large Numbers Are Penciled in to Make 3 × 3 Magic Square

Albert simply split up the 1 through 81 sequence into nine equal units and mentally numbered them as follows:

first unit: 1–9
second unit: 10–18
third unit: 19–27

fourth unit: 28–36
fifth unit: 37–45
sixth unit: 46–54

seventh unit: 55–63
eighth unit: 64–72
ninth unit: 73–81

Then, considering each unit as a single number, Albert placed the first unit, 1–9 (as if he were making a 3 × 3 magic square), in the 3 × 3 space above where he had penciled in the number 1. Next, he placed the second unit, 10–18, in the space penciled 2, and so forth. His progression is shown below, and his completed 9 × 9 on the preceding page.

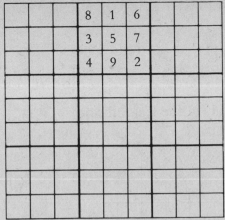

First Move

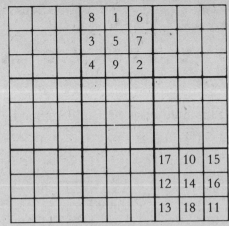

Second Move

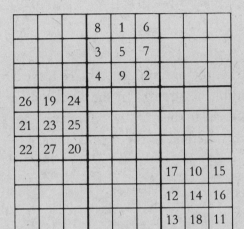

Third Move

Fourth Move

Fifth Move

Method B:
THE PYRAMID METHOD

This method is best understood by a brief study of the diagrams below:

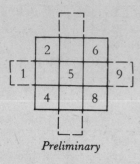

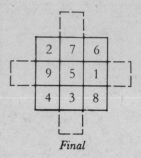

Preliminary   *Final*

In the above arrangement, the numbers outside the preliminary square leave a corresponding number of unfilled boxes within the square. In the final magic square, the outside numbers have been transferred to the unfilled boxes opposite their original location. The application of this method to the 5 × 5 square is shown below.

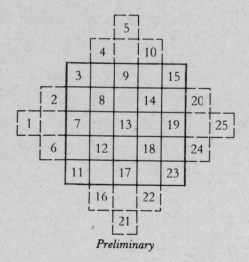

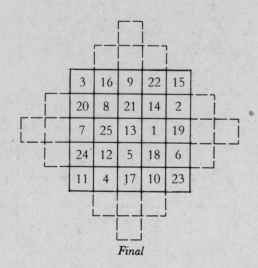

Preliminary   *Final*

Method C: Variation of the Staircase Method

This method is one of the many variations based upon the staircase system of advancing the numbers upward diagonally to the right, as described in Method A. Method A should be practiced and thoroughly understood before attempting this method.

Rule one: Place the number 1 in the box horizontally to the right of the center box.

Rule two: Fill in the boxes by advancing diagonally upward to the right until blocked by a previously placed number.

Rule three (the blocked move): When a move is blocked, place the next number of the series two boxes horizontally to the right of last number placed (see below). The blocked move will occur at every fifth move in a 5 × 5 square, every seventh move in a 7 × 7 square, and so forth. When blocked move occurs, use the concept of imaginary squares as shown in method A.

Arrows indicate blocked moves.

3				
				2
			1	
		5 →	6	
	4			

First Blocked Move

3		9		
	8			2
7			1	
		5		6
11	4		10	

11 (imaginary, to the right)

Second Blocked Move

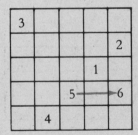

3	16	9		15
	8		14	2
7		13	1	
	12	5		6
11	4		10	

16 (imaginary, to the right)

Third Blocked Move

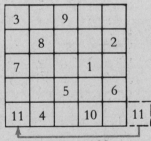

3	16	9	22	15
20	8 → 21	14	2
7	25	13	1	19
24	12	5	18	6
11	4	17	10	23

Fourth Blocked Move

Method d:
THE KNIGHT'S MOVE METHOD

This very interesting system is somewhat similar to method A (the staircase method), and also requires the use of the concept of imaginary squares. It involves the knight's move as in chess. To those unfamiliar with chess, the diagram below depicts the eight moves a knight can make.

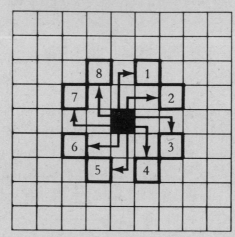

The Eight Knight's Moves

1. 2 up, 1 right   5. 2 down, 1 left
2. 1 up, 2 right   6. 1 down, 2 left
3. 2 right, 1 down   7. 2 left, 1 up
4. 1 right, 2 down   8. 1 left, 2 up

The blocked move is 1 down.

Magic squares may be constructed using all of the eight knight's moves, *depending on the placement of the number 1.*

It should be interesting to a novice magic-square freak who is also a chess bum to find answers to the whys suggested by the foregoing statement.

Suggestions: Place 1 in the middle box of the top row, and using each of the eight knight's moves, see how many original magic squares may be made. Try the 1 somewhere else and see what happens.

In the 5 × 5 magic square below, the number 1 was placed in the middle box of the top row, and the knight's move employed in the construction of the magic square is two boxes up and one to the right. The blocked move is one box down. Note the use of imaginary squares as used in method A (the staircase method).

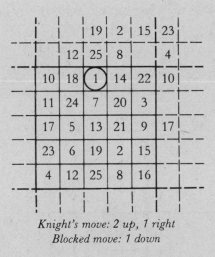

Knight's move: 2 up, 1 right
Blocked move: 1 down

Two knight's moves are shown below with 1 in different positions.

5	13	21	9	17
6	19	2	15	23
12	25	8	16	4
18	(1)	14	22	10
24	7	20	3	11

2 Up, 1 Right

12	3	19	10	21
18	9	25	11	2
24	15	(1)	17	8
5	16	7	23	14
6	22	13	4	20

1 Up, 2 Right

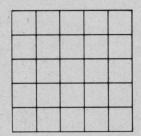 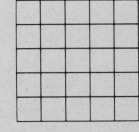

It is also possible to devise expanded knight's moves for odd-order squares larger than 5 × 5.

A DIFFERENT WAY OF CONSTRUCTING KNIGHT'S MOVE SQUARES

1. Place 1 in the middle box of top row.
2. Place the complementary of 1 in the box diametrically equidistant from the center. (In 5 × 5 squares, the complementary to 1 is 25.)
3. Place 2 according to knight's move rules (method D).
4. Place the complementary of 2 in the box diametrically equidistant from center.
5. Place 3 according to knight's move rules (method D).
6. Place the complementary of 3 in the box diametrically equidistant from center.
7. Follow above procedure until magic square is completed. (see 5 × 5 below)

The knight's move is 2 down, 1 to right.

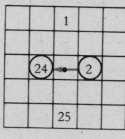

1

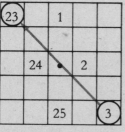

2

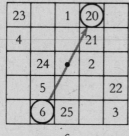

3

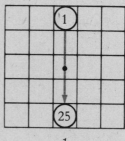

4

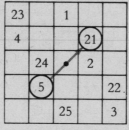

5

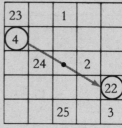

6

SIX VARIATIONS OF THE KNIGHT'S MOVE SHOWING SEQUENCE DESIGNS FOR COMPARISON

The number 1 has been placed in the middle box of the top row of each of the variations shown below. The placement of the 2 (which we have circled in the diagrams) sets the pattern for all subsequent moves within that particular square. For example, if 2 is placed two boxes down and one box to the right of 1, then all subsequent moves within this square must follow this pattern. The blocked move for each of the variations is one down. The six 5 × 5 squares below are completed as far as the placement of 7. Partial sequence designs of each are shown to aid in comparing the different squares.

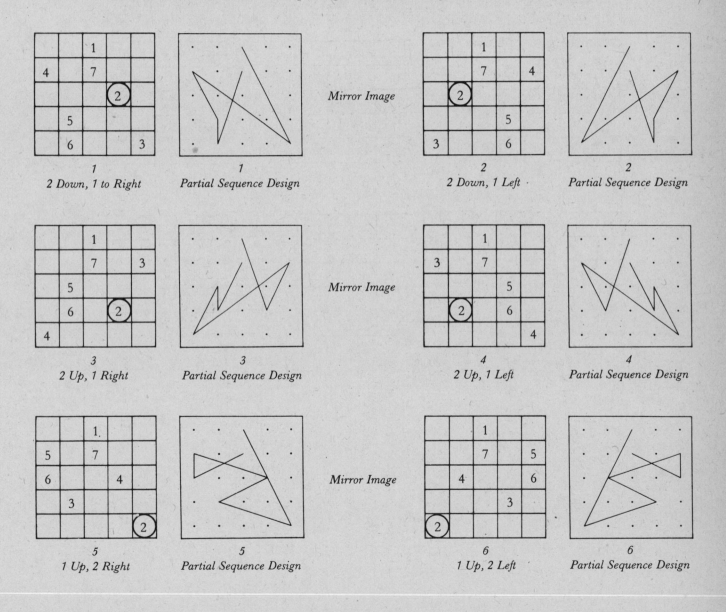

| 1 | 1 |
| 2 Down, 1 to Right | Partial Sequence Design |

Mirror Image

| 2 | 2 |
| 2 Down, 1 Left | Partial Sequence Design |

| 3 | 3 |
| 2 Up, 1 Right | Partial Sequence Design |

Mirror Image

| 4 | 4 |
| 2 Up, 1 Left | Partial Sequence Design |

| 5 | 5 |
| 1 Up, 2 Right | Partial Sequence Design |

Mirror Image

| 6 | 6 |
| 1 Up, 2 Left | Partial Sequence Design |

Method E:
DIFFERENT BLOCKED MOVES

As in the staircase method, all numbers are advanced upward to the right. The blocked moves will vary, as will be shown.

Make a 5 × 5 diagram (or any other odd-order diagram) and place 1 in any box except those forming the diagonal running from the lower left to the upper right corner.

Place 25 (or the last number of the series) in the box diametrically opposite from the 1 (equidistant from center).

The relative spacing between the *last* number and the *first (not the reverse)* determines the blocked moves.

If 1 is placed in the box to the right of the center and 25 is diametrically opposite in the box to the left of the center, it will be seen that 1 is two boxes to the right of 25. The blocked move will thus be two boxes to the right.

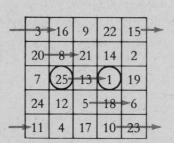

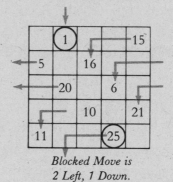

First and Last Numbers Circled.
Blocked Move is Two to Right, as
Shown by Circles.

Blocked Move is
2 Left, 1 Down.

Method F:
ONE OF THE EARLIEST METHODS

The eleventh edition of the *Encyclopaedia Britannica* (1910–1911) refers to this method as "one of the earliest methods used to construct odd-order magic squares." It requires two preliminary squares to construct the third square, which is magic (see below).

3	4	2	5	1
1	**3**	4	2	5
5	1	**3**	4	2
2	5	1	**3**	4
4	2	5	1	**3**

Preliminary Square A
Uses Series 1,2,3,4,5

3	4	0	1	**2**
4	0	1	**2**	3
0	1	**2**	3	4
1	**2**	3	4	0
2	3	4	0	1

Preliminary Square B
Uses Series 0,1,2,3,4

18	24	2	10	11
21	3	9	12	20
5	6	13	19	22
7	15	16	23	4
14	17	25	1	8

Magic Square C

The preliminary square A is constructed by using the series 1,2,3,4,5. Each number is used *five times* and so arranged that no single row or column contains duplicate numbers.

To construct preliminary square A:

1. Place the five 3's (3 is the middle number of the series 1,2,3,4,5) in the corner-to-corner diagonal running from upper left to lower right.
2. Fill in the middle row with the four remaining numbers of the series (1,2,4,5) *in any order.*
3. Fill in the remaining four rows (each of which already contains a 3) with the numbers in the *same relative order* as the middle row. This may be clarified by observing the completed preliminary square A shown above. Note that the 3 in the middle row is followed by the 4 and 2 and preceded by the 5 and 1. The series of the five numbers may be thought of as being strung equally spaced around a charm bracelet. If we start counting with the 3 and go all the way around the bracelet, we find that the order of the numbers is 3,4,2,5,1.

In filling in the remaining four rows of the square, we must keep the relative order of the numbers in each row the same as the middle row. This may best be done by beginning the circular count with the already placed 3 in the row being filled.

Another way to better understand how to keep the numbers in each row in relative order is to observe the order of the numbers in the top row of preliminary square A, which read 3,4,2,5,1.

Instructions for constructing preliminary square A:

1. Make a blank square and fill in the top row with the numbers 3,4,2,5,1.
2. Shift all the numbers of the top row diagonally downward to the right into row two. Note that the number 1 has been bounced out of the square and left an unfilled box at the beginning of row two.
3. Pick up the 1 and place it in the unfilled box at the beginning of row two. Row two now reads 1,3,4,2,5.
4. Shift all numbers of row two diagonally downward to the right into row three. Note that 5 has been bounced out of the square and left an unfilled box at the beginning of row three.
5. Pick up the 5 and place it in the unfilled box of row three, which now will read 5,1,3,4,2.
6. Continue this procedure until all five rows are completed (see below).

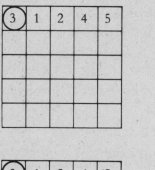

The relative order of the numbers in each row of preliminary square B may be established in the same manner but in reversed order.

Preliminary square B uses the series 0,1,2,3,4. Each number is used five times and so arranged that no single row or column contains a duplicate number.

Instructions for constructing preliminary square B:

1. Place the five 2's (two is the middle number of the 0,1,2,3,4 series) in the corner-to-corner diagonal running from the lower left to the upper right.
2. Fill in the remaining boxes of the middle row with the four remaining numbers of the series (0,1,3,4) *in any order.*
3. Fill in the four remaining rows with the numbers in the same relative positions as they were placed in the middle row.

To construct magic square C:

1. Multiply the number in the upper left corner box of preliminary square B by 5 and add it to the number in the similarly located (upper left corner) box of preliminary square A ($3 \times 5 = 15 + 3 = 18$). Place the total (18) in the similarly located box in square C.
2. Multiply the number in box two of the top row of preliminary square B by 5 and add it to the number in the similarly located box in preliminary square A ($4 \times 5 = 20 + 4 = 24$). Place the total (24) in the similarly located box in square C.
3. The same process is used to fill the remaining boxes and complete the magic square C.

Method F may be used to construct larger magic squares (see 7 × 7 below). The 7 × 7 preliminary square A uses the sequence 1,2,3,④,5,6,7. Square B uses 0,1,2,③,4,5,6. The details of construction are the same as used for the 5 × 5 above except that (because this is a 7 × 7) each number of square B must be multiplied by 7.

4	1	2	6	7	5	3
3	**4**	1	2	6	7	5
5	3	**4**	1	2	6	7
7	5	3	**4**	1	2	6
6	7	5	3	**4**	1	2
2	6	7	5	3	**4**	1
1	2	6	7	5	3	**4**

Preliminary Square A
Using Series 1,2,3,④ *5,6,7.

0	1	6	4	5	2	**3**
1	6	4	5	2	**3**	0
6	4	5	2	**3**	0	1
4	5	2	**3**	0	1	6
5	2	**3**	0	1	6	4
2	**3**	0	1	6	4	5
3	0	1	6	4	5	2

Preliminary Square B
Using Series 0 1 2 ③ 4 5 6

4	8	44	34	42	19	24
10	46	29	37	20	28	5
47	31	39	15	23	6	14
35	40	17	25	1	9	48
41	21	26	3	11	43	30
16	27	7	12	45	32	36
22	2	13	49	33	38	18

Magic Square C

*Circle indicates middle number.

Method G:
Invented by de la Hire

Philippe de la Hire invented this method which he wrote about in his book *Mémoires de l'Académie,* which was published in France in 1705. It is similar to but a definite improvement on the earlier traditional method of using the two preliminary squares to construct a magic square.

3	5	2	1	4
4	**3**	5	2	1
1	4	**3**	5	2
2	1	4	**3**	5
5	2	1	4	**3**

Preliminary Square A
Using Series 1,2,3,4,5

20	5	0	15	**10**
5	0	15	**10**	20
0	15	**10**	20	5
15	**10**	20	5	0
10	20	5	0	15

Preliminary Square B
Using Series 0,5,10,15,20

23	10	2	16	14
9	3	20	12	21
1	19	13	25	7
17	11	24	8	5
15	22	6	4	18

Final
Magic Square C

Preliminary square A is constructed by using the series 1,2,3,4,5. Each number is used *five times,* and numbers are arranged so that no single row or column contains duplicates. Construction details are identical to construction details of preliminary square A in the previous method (F).

Preliminary square B is constructed by using de la Hire's new sequence 0,5,-10,15,20. Each number is used five times, and numbers are arranged so that no single row or column contains duplicate numbers.

To construct preliminary square B:

1. Place the five 10's (10 is the middle number of the 0,5,10,15,20 series) in the corner-to-corner diagonal running from the lower left to the upper right.
2. Fill in the remaining four boxes of the middle row with the four remaining numbers of the series (0,5,15,20) *in any order.*
3. Fill in the four remaining rows (each of which contains a 10) with the numbers in the *same relative order* as the middle row. Arranging the numbers in the same relative order is explained in the construction details of preliminary square A in method F.

To construct magic square C:

Add up the numbers in similarly placed boxes in squares A and B and place the total in similarly placed boxes in magic square C. For example:

Note that the upper left corner box of square A contains the number 3.

Note that the upper left corner box of square B contains the number 20. These two numbers are added and the total (23) is placed in the upper left corner box of square C. Repeat this process to complete the magic square C.

This method may be applied to any larger size odd-order squares (see below).

4	1	2	3	5	6	7
7	4	1	2	3	5	6
6	7	4	1	2	3	5
5	6	7	4	1	2	3
3	5	6	7	4	1	2
2	3	5	6	7	4	1
1	2	3	5	6	7	4

Preliminary Square A
Using Series 1,2,3,(4),5,6,7
7 × 7

14	7	0	28	35	42	21
7	0	28	35	42	21	14
0	28	35	42	21	14	7
28	35	42	21	14	7	0
35	42	21	14	7	0	28
42	21	14	7	0	28	35
21	14	7	0	28	35	42

Preliminary Square B
Using Series 0,7,14,(21),28,35,42
7 × 7

18	8	2	31	40	48	28
14	4	29	37	45	26	20
6	35	39	43	23	17	12
33	41	49	25	15	9	3
38	47	27	21	11	1	30
44	24	19	13	7	32	36
22	16	10	5	34	42	46

Final Magic Square C
7 × 7

Suggestion: Complete final 9 × 9 magic square based on 9 × 9 preliminary squares A and B below:

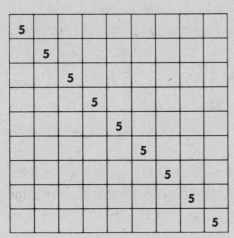

Preliminary Square A
Using Series 1,2,3,4,(5),6,7,8,9
9 × 9

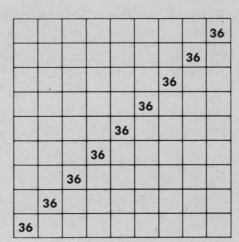

Preliminary Square B
Using Series 0,9,18,27,(36),45,54,63,72
9 × 9

METHOD H: TRANSFORMATION BY INTERCHANGE OF QUADRANTS

AS EXPLAINED BY MAURICE KRAITCHIK

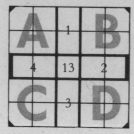

Diagram X

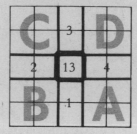

Diagram Y

8	17	24	5	11
16	23	2	14	10
25	1	13	7	19
4	15	6	18	22
12	9	20	21	3

Magic Square X

18	22	6	4	15
21	3	20	12	9
7	19	13	25	1
5	11	24	8	17
14	10	2	16	23

New Magic Square Y

Diagrams X and Y, when applied to magic square X and Y, show the interchange of quadrants necessary to transform magic square X into the new magic square Y.

In comparing diagram Y with diagram X, note that each labeled section in diagram X has been moved to the opposite position in diagram Y. (Section A has exchanged positions with section C, section 1 has exchanged positions with section 3, and so forth.)

The same exchange pattern is used to transform the magic square X into the new magic square Y.

Is another interchange system possible?

Method I:
MARTIN GARDNER'S

Method I as shown by Martin Gardner in *Scientific American* (January 1976) will produce other magic squares but *only* when the initial original magic square is associative and pandiagonal (see the glossary).

A. Construct associative-pandiagonal original magic square.

1	15	24	8	17
23	7	16	5	14
20	4	13	22	6
12	21	10	19	3
9	18	2	11	25

Original Associative and Pandiagonal
Magic Square

This square made by modified method D (knight's move).
1 Up, 2 Right
Block Move: 1 Down, 1 Right

B. Repeat the above square in an infinite pattern as shown below.

1	15	24	8	17	1	15	24	8	17	1	15	24	8	17
23	7	16	5	14	23	7	16	5	14	23	7	16	5	14
20	4	13	22	6	20	4	13	22	6	20	4	13	22	6
12	21	10	19	3	12	21	10	19	3	12	21	10	19	3
9	18	2	11	25	9	18	2	11	25	9	18	2	11	25
1	15	24	8	17	1	15	24	8	17	1	15	24	8	17
23	7	16	5	14	23	7	16	5	14	23	7	16	5	14
20	4	13	22	6	20	4	13	22	6	20	4	13	22	6
12	21	10	19	3	12	21	10	19	3	12	21	10	19	3
9	18	2	11	25	9	18	2	11	25	9	18	2	11	25

C. We can outline a 5 × 5 square anywhere on this plane and it will be magic, though not necessarily associative. To be associative, it must have 13 in the center.

METHOD J: EXCHANGE OF BORDER COLUMNS AND ROWS

A. Begin with any odd-order magic square (see below).

8	17	24	5	11
16	23	2	14	10
25	1	13	7	19
4	15	6	18	22
12	9	20	21	3

A

B. Exchange left and right border columns.

11	17	24	5	8
10	23	2	14	16
19	1	13	7	25
22	15	6	18	4
3	9	20	21	12

B

C. Exchange top and bottom border rows of square B.

3	9	20	21	12
10	23	2	14	16
19	1	13	7	25
22	15	6	18	4
11	17	24	5	8

C
Completed Magic Square

Method K: Exchange of Two Rows and Two Columns

A. Begin with any odd-order magic square.

8	17	24	5	11
16	23	2	14	10
25	1	13	7	19
4	15	6	18	22
12	9	20	21	3

A

B. Exchange rows 1 and 2 and rows 4 and 5.

16	23	2	14	10
8	17	24	5	11
25	1	13	7	19
12	9	20	21	3
4	15	6	18	22

B

C. Exchange columns 1 and 2 and columns 4 and 5 of square B.

23	16	2	10	14
17	8	24	11	5
1	25	13	19	7
9	12	20	3	21
15	4	6	22	18

C
Completed Magic Square

METHOD L: THE BASIC GEOMETRIC DESIGN METHOD

INVENTED BY L. S. FRIERSON

This appears to be an improvement on de la Hire's method (method G) of using two preliminary and one final square. Freak Frierson has succeeded in opening up unique ways of making a large number of original squares with this method.

5 × 5 squares will be used to explain the system, which (like the previous methods shown) can be applied to odd-order squares of any size.

1. The series 1,2,3,4,5 is used in preliminary square A, and each number is used five times. The construction of preliminary square A is begun by placing 3 (the middle number of the series) in the center box and placing the remaining two pairs of 3's equidistant from the center, arranged so that each column and each row contains but one 3 (see square A below).

2. The series 0,5,10,15,20 is used in preliminary square B, and each number is used five times. Construction of B is begun by placing 10 (the middle number of the series) in the center box and placing the remaining two pairs of 10's in reversed positions in relation to the placement of 3's in square A (see below).

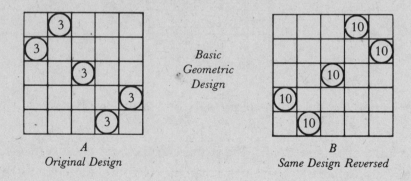

Basic Geometric Design

A
Original Design

B
Same Design Reversed

3. Fill in the middle row of square A with the remaining numbers of the 1 through 5 series arranged in any way *provided* that the pairs of complementary numbers (see glossary) are equidistant from the center. In this instance, the complementary pairs are 1 and 5, 2 and 4 (see square A below).

4. Fill in the middle row of square B with the remaining numbers of the 0,5,10,15,20 series arranged in any way *provided* that the pairs of complementary numbers are equidistant from the center. In this instance, the complementary pairs of numbers are 0 and 20, 5 and 15 (see square B below).

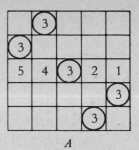

A

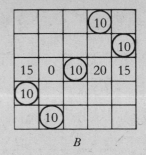

B

Note that the middle row of square A above could have been arranged in several ways other than the one selected. For example, 4,5,③,1,2; 2,1,③,5,4; 5,2,③,4,1; etc.

Note that the middle row of square B could also have been arranged in different ways. For example, 0,5, ⑩ ,15,20; 0,15, ⑩ ,5,20; 5,0, ⑩ ,20,15; etc.

5. Fill in the remaining rows of square A, placing the numbers in the *same relative order* as they have been placed in the middle row. (As explained in method F.)

6. Fill in the remaining rows of square B in the same manner as above.

7. The magic square C is completed in the same manner as explained in method G. See completed A, B, and C squares below.

4	③	2	1	5
③	2	1	5	4
5	4	③	2	1
2	1	5	4	③
1	5	4	③	2

A

5	15	0	⑩	20
20	5	15	0	⑩
15	0	⑩	20	5
⑩	20	5	15	0
0	⑩	20	5	15

B

9	18	2	11	25
23	7	16	5	14
20	4	13	22	6
12	21	10	19	3
1	15	24	8	17

C

The advantage of this method lies not only in the number of different magic squares made possible by different arrangements of the numbers in the middle rows of squares A and B, but also in the variety made possible by the use of different basic geometric designs, one of which is shown below.

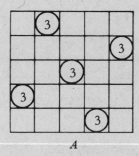

A

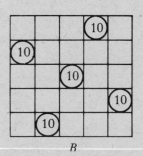

B

There are other basic geometric designs for the 5 × 5. Can you figure them out?

Two basic geometric designs for the 7 × 7 square are shown below.

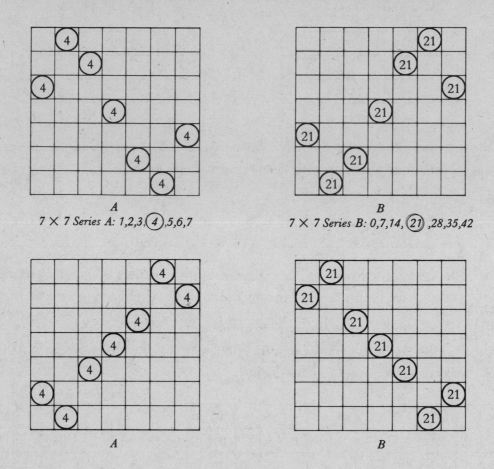

A

7 × 7 Series A: 1,2,3,④,5,6,7

B

7 × 7 Series B: 0,7,14,㉑,28,35,42

A

B

There are also other arrangements within the 7 × 7. Can you figure them out?

Two basic geometric designs for the 9 × 9 square are shown below.

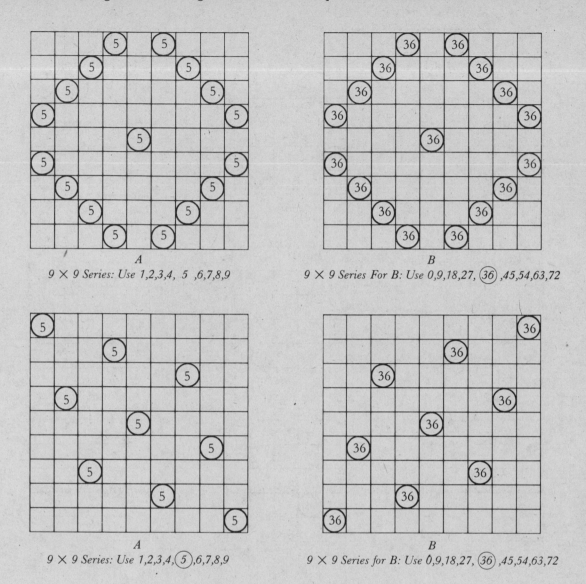

A

9 × 9 Series: Use 1,2,3,4, 5 ,6,7,8,9

B

9 × 9 Series For B: Use 0,9,18,27, (36) ,45,54,63,72

A

9 × 9 Series: Use 1,2,3,4,(5),6,7,8,9

B

9 × 9 Series for B: Use 0,9,18,27, (36) ,45,54,63,72

Frierson says there are other basic geometric designs for the 9 × 9. Can you figure them out?

In some instances it is possible to construct original magic squares by using *two entirely different* basic geometric designs in preliminary A and B squares (see below).

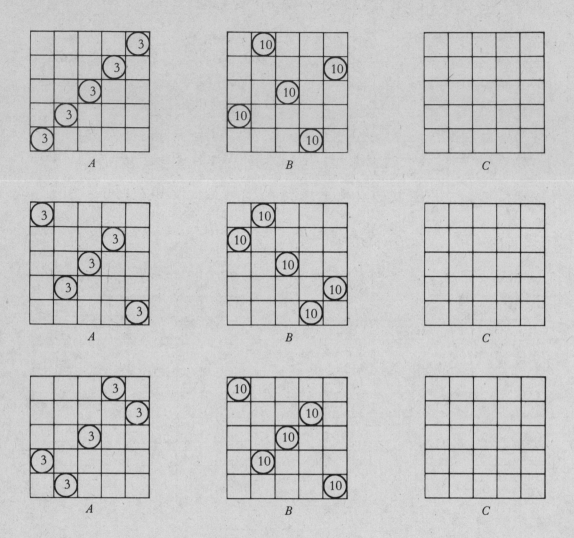

Suggestions: Complete the above squares. Can you find other combinations that will work?

Method M: Figures
of Equilibrium Method

INVENTED BY L. S. FRIERSON

Freak Frierson does it again! Not content to stretch out on his bags of fan mail and rest on his laurels after giving birth to method L, Brother Frierson has knocked himself out and come up with one of the wildest schemes yet.

You will recall that in his previous method (L), Frierson introduced the super-idea of the basic geometric designs for preliminary squares A and B. He now goes even farther out and has hatched a new breed of A and B preliminary squares based on what he calls figures of equilibrium. Here again he sets sail into uncharted magic-square seas.

The 5 × 5 preliminary square A (see below) shows a diagram of a figure of equilibrium. If the circles in this square were to represent equal weights, connected as shown by the dotted lines, the weights would balance evenly if the square were suspended from the center. Frierson says there are five different figures of equilibrium for the 5 × 5 square. (I'll believe this when I see them.)

Like method L, this method uses two preliminary squares, A and B. Square A uses sequence 1,2,3,4,5. Square B uses sequence 0,5,10,15,20.

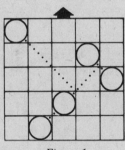

Figure 1
Preliminary Square A

Instructions for constructing preliminary square A, using series 1,2,3,4,5:

1. Referring to the design in figure 1, make a blank 5 × 5 square and fill in the five circles with the number 1. Place directional arrow at top (see below).

Figure 2

2. With the other numbers of the series (except the middle number, 3), make three other identical figures of equilibrium, but turn each within the square in the direction indicated by the arrow. The result is shown separately in figures 3, 4, and 5, and collectively in figure 6.

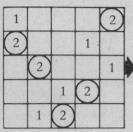

Figure 3
Design Turned to Right (Note Arrow)
2 Placed in Each Circle

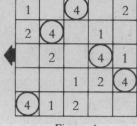

Figure 4
Design Turned to Left (Note Arrow)
4 Placed in Each Circle

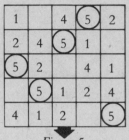

Figure 5
Design Turned Upside Down (Note Arrow)
5 Placed in Each Circle

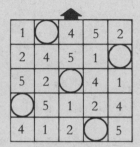

Figure 6
Preliminary Square A Complete Except for
Five Remaining Unfilled Boxes

3. The five remaining unfilled boxes (Figure 6) will be geometrically balanced and must be filled with the middle number of the series (3), thereby completing preliminary square A, as shown in figure 7 below.

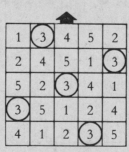

Figure 7

Most magic-square freaks (including the author) may find it difficult to construct preliminary square A even after understanding the step-by-step process as shown above, since it necessitates turning the figure of equilibrium to a different position within the square each time a new number of the series is to be placed.

An easier way to construct preliminary square A:

1. Make a 5 × 5 diagram as shown in figure 1 above and place the number 1 in each of the circles.
2. Using a separate piece of paper, make a duplicate diagram, cut holes where the five circles are, and place this cutout over the original diagram so that the five 1's show through the holes.
3. Draw directional arrow on the cutout pointing up.
4. Place cutout over the square, turn arrow to the right (as shown in figure 3), and write the number 2 through the five cutout holes onto the square, which will now look like figure 3.
5. Turn cutout to left (as shown in figure 4) and write in the five 4's.
6. Turn cutout to the down position (as shown in figure 5) and write in the five 5's.
7. Remove the cutout and fill in the five remaining unfilled boxes with the number 3, completing the construction of preliminary square A as shown in figure 7.

Instructions for constructing preliminary square B with method M, using series 0,5,10,15,20:

Use the same paper cutout in same manner as used in constructing square A, but with the following variations:

1. Turn cutout over with back side up and place arrow in same position as reverse side.
2. Place on top of blank B square diagram, turn arrow to the down position and write 0 in each of the five circles.
3. Turn arrow to left position and write in the 5's.
4. Turn arrow to right position and write in the 15's.
5. Turn arrow to up position and write in the 20's.
6. Write 10 in the remaining five boxes to complete preliminary square B (see below).

To form magic square C, combine preliminary squares A and B as shown in method G. See the completed square C below.

1	3	4	5	2
2	4	5	1	3
5	2	3	4	1
3	5	1	2	4
4	1	2	3	5

Preliminary Square A

5	0	15	10	20
10	20	0	15	5
20	15	10	5	0
15	5	20	0	10
0	10	5	20	15

Preliminary Square B

6	3	19	15	22
12	24	5	16	8
25	17	13	9	1
18	10	21	2	14
4	11	7	23	20

Magic Square C

The great difference between this and other methods may be noted by observing that, while the usual complementary pairs of numbers straddle the central number 3 in preliminary square A, *the relative order of the numbers in each row is different.* This also holds true for square B. For this reason it is supposed that magic squares constructed by this method are unique and may not be constructed by any other method.

METHOD N: THE SUPER-GEOMETRIC DESIGN METHOD

Author's note: My good friend Herb the Falconer lives far out in the country, which prevents my seeing him as much as I would like. I began writing him about magic squares over two years ago when I first got hooked. His math education is about on a par with mine, which isn't saying much, and he is at a disadvantage compared to me in that research books are not readily available to him.

Through our regular correspondence, nevertheless, he has become a magic-square freak (1st class) and is especially interested in 5 × 5 squares.

Herb is familiar with the first portion of this book dealing with odd-order squares and has made many valuable observations and suggestions.

Here is one of my many letters to Herb, which I hope might prove informative to the reader at this point.

Dear Herb:

I thought I had pretty well covered the various methods of constructing odd-order magic squares in my earlier letters. (Silly boy! I should have known better.) I am now of the opinion that this subject will probably never be thoroughly covered. Every time I begin to feel that I know something, a brand-new revelation comes skidding in from left field and makes me realize that although "I've come a long way from St. Louis," I still have a long way to go—but this only turns me on all the more.

Anyhow, just to make sure I hadn't overlooked anything important about odd-order magic squares that should be included in my book, I went back over the highly condensed article in the 1911 edition of the *Encyclopaedia Britannica* and noticed three squares (two preliminary and the final magic) attributed to de la Hire. The squares were shown but no construction method was explained. Also the squares were so tiny that the numbers were barely legible. It's a good thing I looked closely, as you'll see later.

At first glance, I thought these squares were made by the same de la Hire method (method G) I outlined to you some time ago. They do have similarities in that the usual 1,2,3,4,5 series in square A is combined with the usual 0,5,-10,15,20 series in square B to form the magic square.

2	1	5	3	4
3	4	2	1	5
1	5	3	4	2
4	2	1	5	3
5	3	4	2	1

Preliminary Square A
Series 1,2,3,4,5

15	5	0	20	10
0	20	10	15	5
10	15	5	0	20
5	0	20	10	15
20	10	15	5	0

Preliminary Square B
Series 0,5,10,15,20

17	6	5	23	14
3	24	12	16	10
11	20	8	4	22
9	2	21	15	18
25	13	19	7	1

Final Magic Square

In looking closer at the B square, I saw something that struck me as quite strange. *5 was in the center box!* In previous methods, only the middle number of the series is permitted in the center box. 5 was sitting where 10 usually sits. This is heresy—*very* peculiar.

Being curious to see if the duplicate numbers in squares A and B formed any kind of a revealing pattern, I drew a column of five separate A squares and beside it a second column of five separate B squares, as you may see on the opposite page.

In the top square of the first column, I have placed the five 1's in the same positions shown in de la Hire's square A.

Moving downward, I have placed the five 5's, then the five 3's, etc. (Notice that the order of these numbers corresponds to the order of the numbers in the middle row of de la Hire's square A.)

The column of five B squares was made the same way.

Note that the third or middle square of each of the two columns reveals a design used by Frank Frierson in his methods. (The design is reversed in the B square.) I have emphasized this design with heavy lines.

By the way, if you have forgotten Frierson's methods, refresh your memory by going back over my letters; otherwise you won't be able to appreciate the magnitude of what I'm about to lay on you. You will recall that in both of Frierson's methods, the numbers used in the center box of this design were *always* 3 or 10—no deviation; that is, the middle numbers of the A or B series.

I started to investigate what the 5 was doing in the center box of de la Hire's square B and E U R E K A !

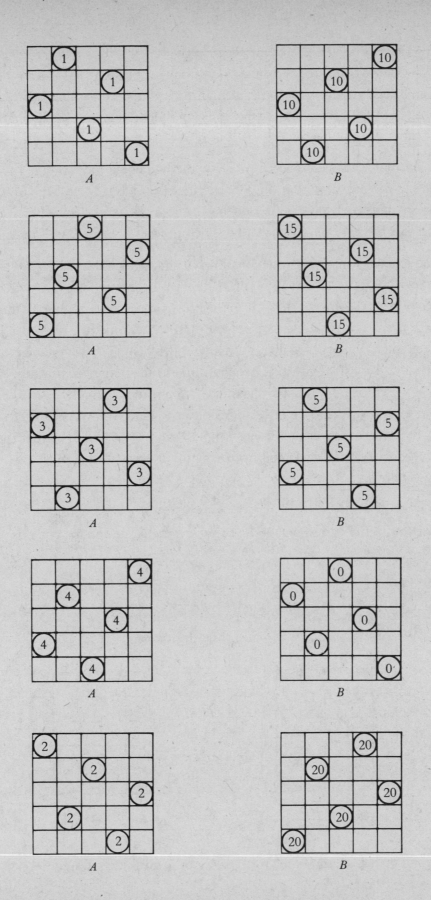

I think I have discovered a brand-new method by which a vast number of original 5 × 5 magic squares may be made. (I figure at least 14,400—maybe more.) I haven't seen this method described in any of the research books I've been studying, but maybe it was there and concealed from me by indecipherable mathematical terminology.

To claim that I have discovered this new method doesn't mean that I think I'm the *first* to discover it—after all, Christopher Columbus wasn't the first man to discover America. I have no doubt that some freak several hundred years ago figured this out all by himself, and I'll bet he was as excited as I am.

I am so excited I'm giving a major blowout this weekend to celebrate the great discovery I'm about to reveal to you. I hope you can attend. You were here for the party I gave when I finally discovered *on my own* (I hadn't done any research then) how to make a 4 × 4. But I don't think you really understood my excitement and I'm going to tell you why. You never stayed up until 3:00 A.M. and rassled with this problem like I did. I *showed* you how to make the 4 × 4. You said, "That's very nice." "Very nice," my eye! *It was terrific!* Of course, it's very simple once somebody shows you how to do it. After all, there are only sixteen little numbers and all you've got to do is to get them into the sixteen boxes so they'll add up the same in every direction. But it's not so simple if you start from scratch. Try this out on some of your friends out there in Falcon country. Here's the new (old?) method:

I choose to call the design below the super-geometric design for obvious reasons, as you shall see.

As I said before, Frierson used this design in both his basic geometric and figures of equilibrium methods, but he didn't mention that there was anything special about it.

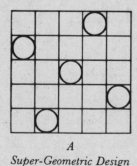

A
Super-Geometric Design

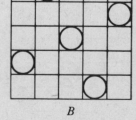

B
Super-Geometric Design (Reversed)

Get a load of *this* method and you'll see why I think this design deserves to be called super.

1. Lay out super-geometric design as in square A above.
2. Place *any one number* of the 1,2,3,4,5 series in each of the five circles of the super-geometric design.
3. Complete the middle row by placing the remaining four numbers of the series *in any order. They don't have to be complementary!*
4. Complete the remaining four rows by placing the numbers in the same relative order as the middle row.
5. Lay out super-geometric design as in square B above.
6. Place any one number of the 0,5,10,15,20 series in each of the five circles of the super-geometric design.
7. Complete the middle row by placing the remaining four numbers of the series *in any order.*
8. Complete square B by filling in the remaining four rows with the numbers in the same relative order as arranged in middle row.
9. Combine square A and square B in the usual fashion to complete magic square C. See two examples of this method below.

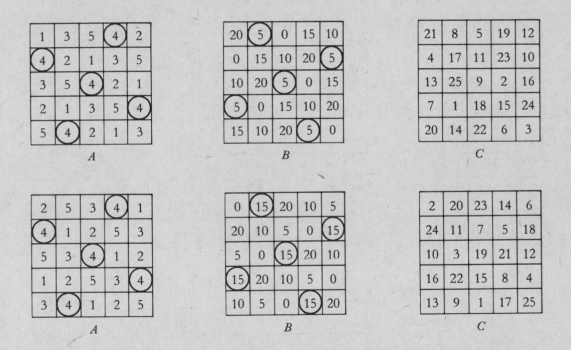

How does this grab you? Happy hawking.

As ever,
Jim

Author's note: The day after I put this letter in the mail, I received a letter from Herb the Falconer, enclosing his H-bomb (H for Herb) method for 5 × 5's that follows. Our letters crossed in the mail, and it is interesting that he too saw something special about the super-geometric design.

A great variety of 7 × 7 magic squares may be constructed using a super-geometric design similar to the one shown in the previous 5 × 5 method.

Two examples are shown below:

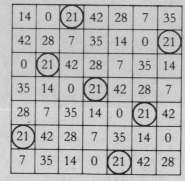

1	7	2	5	(4)	3	6
(4)	3	6	1	7	2	5
6	1	7	2	5	(4)	3
7	2	5	(4)	3	6	1
5	(4)	3	6	1	7	2
3	6	1	7	2	5	(4)
2	5	(4)	3	6	1	7

A
Series 1,2,3,(4),5,6,7
Using Middle Number in Center

14	0	(21)	42	28	7	35
42	28	7	35	14	0	(21)
0	(21)	42	28	7	35	14
35	14	0	(21)	42	28	7
28	7	35	14	0	(21)	42
(21)	42	28	7	35	14	0
7	35	14	0	(21)	42	28

B
Series 0,7,14,(21),28,35,42
Using Middle Number in Center

15	7	23	47	32	10	41
46	31	13	36	21	2	26
6	22	49	30	12	39	17
42	16	5	25	45	34	8
33	11	38	20	1	28	44
24	48	29	14	37	19	4
9	40	18	3	27	43	35

C
Magic Square

3	4	5	1	(6)	7	2
(6)	7	2	3	4	5	1
2	3	4	5	1	(6)	7
4	5	1	(6)	7	2	3
1	(6)	7	2	3	4	5
7	2	3	4	5	1	(6)
5	1	(6)	7	2	3	4

A
Series 1,2,3,4,5,6,7
Not Using Middle Number in Center

35	42	(7)	0	14	21	28
0	14	21	28	35	42	(7)
42	(7)	0	14	21	28	35
28	35	42	(7)	0	14	21
14	21	28	35	42	(7)	0
(7)	0	14	21	28	35	42
21	28	35	42	(7)	0	14

B
Series 0,7,14,21,28,35,42
Not Using Middle Number in Center

38	46	12	1	20	28	30
6	21	23	31	39	47	8
44	10	4	19	22	34	42
32	40	43	13	7	16	24
15	27	35	37	45	11	5
14	2	17	25	33	36	48
26	29	41	49	9	3	18

C
Magic Square

Suggestion: Design your own super-geometric design for 9 × 9 and other larger odd-order squares.

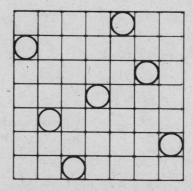

SYMMETRICAL AND ASYMMETRICAL MAGIC SQUARES

A. SYMMETRICAL MAGIC SQUARES

Sequence designs will reveal that all associative and pandiagonal-associative magic squares are symmetrical (see below).

Pandiagonal-associative magic square made by staircase method (A). Note relationship of first, middle, and last numbers of the series (1,13,25).

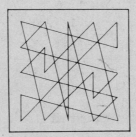

17	24	①	8	15
23	5	7	14	16
4	6	⑬	20	22
10	12	19	21	3
11	18	㉕	2	9

Symmetrical

Pandiagonal-associative magic square made by knight's move method (D). Note relationship of 1,13,25.

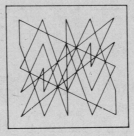

10	18	①	14	22
11	24	7	20	3
17	5	⑬	21	9
23	6	19	2	15
4	12	㉕	8	16

Symmetrical

Associative, but not pandiagonal-associative, made by geometric figures method (M). Note relationship of 1,13,25.

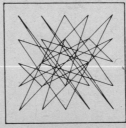

9	18	2	11	㉕
23	7	16	5	14
20	4	⑬	22	6
12	21	10	19	3
①	15	24	8	17

Symmetrical

B. ASYMMETRICAL MAGIC SQUARES

If the first, middle, and last numbers of a magic square do not form a symmetrical relationship, it may be assumed that a sequence design of that square will be asymmetrical.

Question: Are there exceptions to this "rule"?

Magic square made by method F. Note relationship of 1,13,25.

18	24	2	10	11
21	3	9	12	20
5	6	(13)	19	22
7	15	16	23	4
14	17	(25)	(1)	8

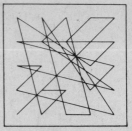

Asymmetrical 1

Magic square made by method G. Note relationship of 1,13,25.

23	10	2	16	14
9	3	20	12	21
(1)	19	(13)	(25)	7
17	11	24	8	5
15	22	6	4	18

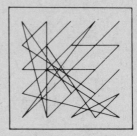

Asymmetrical 2

Magic square. Note relationship of 1,13,25.

17	6	5	23	14
3	24	12	16	10
11	20	8	4	22
9	2	21	15	18
(25)	(13)	19	7	(1)

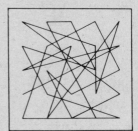

Asymmetrical 3

THE KAHN H–BOMB

(H FOR HERB)

The Kahn preliminary A- and B-bombs, when coupled, produce the crazy, cataclysmic peace-head known as the Kahn H-bomb.

The preliminary A- and B-bombs consist of atoms and molecules as well as a small assortment of nuclear numerical devices, which when combined result in a quickly constructed H-bomb.

To explain and clarify the construction of the H-bomb, it was found necessary to create a new terminology, given in the following glossary.

HERB'S GLOSSARY

ATOM   The smallest particle of an atomic sequence.

ATOMIC GRID   The two atomic sequences of numbers 1,2, 3 ,4,5 and 0,5, 10 ,15,20.

ATOMIC SYMMETRY   The process whereby a molecule generates a nucleus which will spawn a prettier, more beautiful bomb after fission takes place.

CATALYSIS   The change and especially the increase in the rate of numerical reaction brought about by the introduction to the nuclear mass of the catalysts 3 and 10 .

CATALYSTS   The atoms 3 and 10 that will change and increase in the rate of numerical reaction brought about by them upon their respective masses.

ELEMENT   A basic constitutive part of the atomic mass consisting of a pair of numerical atoms.

M.S.F.S.   A magic-square freak scientist.

MOLECULE   The combination of a pair of elements (4 atoms) which form the smallest particle of matter that is the same as the whole of the mass.

NUCLEUS   A central mass about which numerical matter gathers or is collected (the 3 × 3 within a 5 × 5).

NUCLEAR MEMBRANE   A thin, almost invisible graphite structure, basically square, which surrounds the nucleus.

REACTOR A heavy graphite structure in which a numerical chain reaction is initiated and controlled (a 5 × 5 square).

CATALYTIC CONTAINER One of five circular graphite enclosures arranged in a geometric pattern on an atomic grid within a reactor (see below).

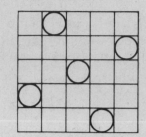

Note: There are no two containers in any row, column, or diagonal.

CONSTRUCTION DETAILS OF THE KAHN A-BOMB

In order to build the magical Kahn H-bomb, we must first construct the Kahn A-bomb. This entails the use of the atomic sequence 1,2, 3 ,4,5. Since the atom 3 will be our catalyst and will be added to the atomic pile later, we put this aside for the moment. We are now left with the atoms 1,2,4,5, which are used to construct 12 pairs of elements. They are:

1 and 2	1 and 4	1 and 5
2 and 1	2 and 4	2 and 5
4 and 1	4 and 2	4 and 5
5 and 1	5 and 2	5 and 4

From the above elements, we may now proceed to build our molecules. It must be stressed that no atoms of the same strength may be used in the initial numerical molecule.

The 24 molecules that can be used are shown on the opposite page.

Note that when a molecule is formed one element takes the north-south position and the other element the east-west position. The inventor mentions this for the M.S.F.S. who might also enjoy playing bridge.

Each of the above molecules consists of four atoms and no atom is duplicated in the molecule.

For ease of identification, each of the above A-bomb molecules has been lettered.

Rule 1: Never duplicate an atom of the same strength in a molecule.

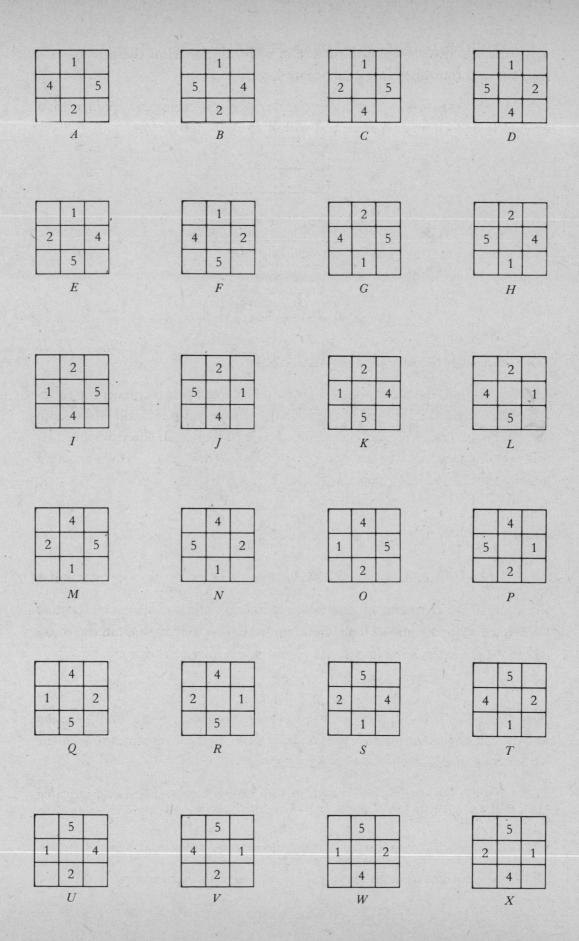

1. Build a nuclear membrane on an atomic grid.

2. Insert one of the 24 molecules within the nuclear membrane. The inventor arbitrarily has selected the J molecule to demonstrate.

	2	
5		1
	4	

3. The lower left corner within the nuclear membrane is very important in the construction of the A-bomb. For identification, we show this corner with a circle.

	2	
5		1
◯	4	

Rule 2: Never duplicate an atom of the same strength in any row, column, or diagonal.

When constructing an A-bomb, the numerical atom that appears in the box to the right of center is the one that is placed in the lower left corner.

	2	
5		1
①	4	

4. We will now fill in the opposite upper right corner with the other atom of the element.

	2	5
5		1
①	4	

5. To complete the nucleus, we must fill in the other two corners with the atoms of the other elements. *Refer to Rule 2.*

Completed Nucleus

6. Erect a reactor upon the atomic grid and insert the nucleus within it.

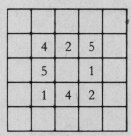

Reactor 5 × 5

7. It is now time to construct the catalytic containers. The A-bomb containers have the following pattern:

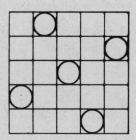

Note: No container is duplicated in any row, column, or diagonal.

8. We now show the A-bomb under construction with the containers in place filled with the catalyst ③.

9. It will now be noticed that the bomb contains two rows and two columns that have four atoms in each. One atom of the atom sequence is missing in each of these rows and columns. Insert these missing atoms.

	3		(4)	
(1)	4	2	5	3
	5	3	1	
3	1	4	2	(5)
	(2)		3	

10. We must now fill in the diagonal corners with the missing atoms. *Be sure* not to duplicate any atom in the same row, column, or diagonal.

(5)	3		4	(2)
1	4	2	5	3
	5	3	1	
3	1	4	2	5
(4)	2		3	(1)

11. You will now note that there is one atom missing again from two rows and two columns. Fill these in but keep rule 2 in mind. Try not to get rattled, as this might upset the atomic order within the A-bomb.

CONSTRUCTION DETAILS OF THE
KAHN B-BOMB

The building of the B-bomb is identical to that of the A-bomb, but utilizes the atomic sequence 0,5, 10 ,15,20. Since our catalyst will be atom 10 , we are left with the atoms 0,5,15,20. The elements that we will use to construct the B-bomb are shown below:

0 and 5	0 and 15	0 and 20
5 and 0	5 and 15	5 and 20
15 and 0	15 and 5	15 and 20
20 and 0	20 and 5	20 and 15

The molecules that can be assembled using the above elements are shown on the opposite page.

For ease of identification, each of the B-bomb molecules has been numbered.

1

	0	
15		20
	5	

2

	0	
20		15
	5	

3

	0	
5		20
	15	

4

	0	
20		5
	15	

5

	0	
5		15
	20	

6

	0	
15		5
	20	

7

	5	
15		20
	0	

8

	5	
20		15
	0	

9

	5	
0		20
	15	

10

	5	
20		0
	15	

11

	5	
0		15
	20	

12

	5	
15		0
	20	

13

	15	
5		20
	0	

14

	15	
20		5
	0	

15

	15	
0		20
	5	

16

	15	
20		0
	5	

17

	15	
0		5
	20	

18

	15	
5		0
	20	

19

	20	
5		15
	0	

20

	20	
15		5
	0	

21

	20	
0		15
	5	

22

	20	
15		0
	5	

23

	20	
0		5
	15	

24

	20	
5		0
	15	

12. Erect a nuclear membrane and insert one of the above molecules. The inventor has selected the molecule 10.

	5	
20		0
○	15	

13. As in step 3, the lower left corner, which has been circled above, is very important. When constructing a B-bomb, the atom that goes in the circled box is the one that appears directly above the center box.

	5	
20		0
⑤	15	

14. Fill in the upper right corner with the other atom of the element, and then, using the other element, fill in the other corners.

0	5	15
20		0
5	15	20

Completed Nucleus

15. Erect a reactor upon the atomic grid and insert the nucleus within it.

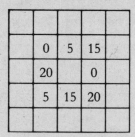

Reactor 5 × 5

16. Construct the catalytic containers, but note that they have a reverse or mirror image of the A containers.

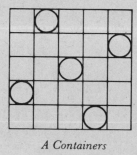

A Containers

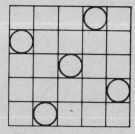

B Containers

17. We now show the B-bomb under construction with the catalytic containers in place and filled with the catalyst ⑩ .

			⑩	
⑩	0	5	15	
	20	⑩	0	
	5	15	20	⑩
	⑩			

18. Proceed now as in the construction of the A-bomb and fill in the missing atoms. When completed, it will look like this:

5	15	20	10	0
10	0	5	15	20
15	20	10	0	5
0	5	15	20	10
20	10	0	5	15

Below, we show the completed A- and B-bombs and have erected a third reactor which will house the all-powerful Kahn H-bomb. Fission will now take place.

5	3	1	4	2
1	4	2	5	3
2	5	3	1	4
3	1	4	2	5
4	2	5	3	1

Bomb A

5	15	20	10	0
10	0	5	15	20
15	20	10	0	5
0	5	15	20	10
20	10	0	5	15

Bomb B

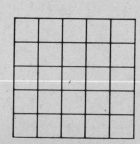

Reactor for H-Bomb

19. Add up the atoms in each similarly located box of the bombs A and B and insert the total in the identical box of the reactor.

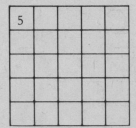   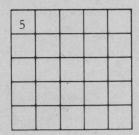   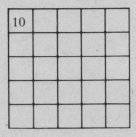

20. Continue this process until the H-bomb is complete. If it has been constructed properly, the totals of each row, column, and diagonal will add up to 65. If it does, then erect another reactor to check if this bomb is symmetrical.

5	3	1	4	2
1	4	2	5	3
2	5	3	1	4
3	1	4	2	5
4	2	5	3	1

A-Bomb

5	15	20	10	0
10	0	5	15	20
15	20	10	0	5
0	5	15	20	10
20	10	0	5	15

B-Bomb

10	18	21	14	2
11	4	7	20	23
17	25	13	1	9
3	6	19	22	15
24	12	5	8	16

H-Bomb

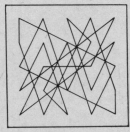

Symmetrical Sequence Design

21. To check if an H-bomb is symmetrical, merely connect with lines the numbers 1 and 2, 2 and 3, and so forth.

In building A- and B-bombs, either of the two arrrangements may be used with molecules A and B.

Both patterns below will result in magic squares.

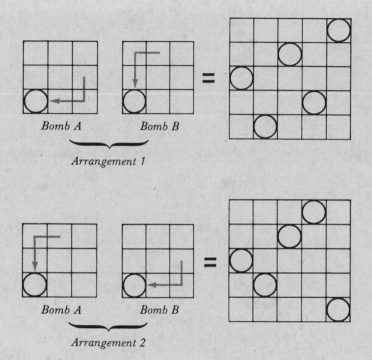

These arrangements will not result in magic squares:

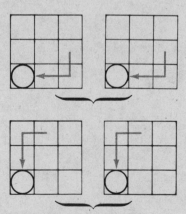

Magic squares and
Pythagorean numbers

The famous picture *Melencolia* engraved by Albrecht Dürer in 1514, which was briefly mentioned in the introduction, gives us reason to infer that magic squares in past centuries held a much deeper meaning in the minds of men than that of simple mathematical curiosities.

Regardless of how we may choose to interpret the obscure overall symbolism of the engraving, it is reasonable to assume that the inclusion of the prominently placed 4 × 4 magic square had some great symbolic significance to the artist.

Although there appears to be no existing evidence of magic squares being known to the ancient Egyptians, it is known that they did arrange numbers in the form of magic diagrams. Some historians believe that Pythagoras (582–507 B.C.), returning to Greece from foreign travels, brought this knowledge with him.

Plato (427–347 B.C.), writing about Pythagorean philosophy and ideas in his *Republic,* makes vague references to the number 729, which is found to be of major importance throughout the Pythagorean system. Plato also quotes Socrates (469–399 B.C.) who, speaking of the number 729, says " . . . a number which closely concerns human life, if human life is concerned with days and nights and months and years." Among other references not specifically stated, Socrates is undoubtedly alluding to the solar year of 365 days and 364 nights (365 + 364 = 729).

Plutarch (46–120 A.D.) states that the number 729 belongs to the sun, 243 to Venus, 81 to Mercury, 27 to the moon, 9 to the earth, and 3 to Antichthon ("the earth opposite ours"). These numbers, of course, are progressive multiples of 3: $3 \times 3 = 9$; $3 \times 9 = 27$; $3 \times 27 = 81$; $3 \times 81 \times 243$; $3 \times 243 = 729$.

The above and many similar numbers were derived from progressions of four numbers such as 1:2::4:8 and 1:3::9:27. They are considered of major importance by the Pythagoreans. Plato combined them into one series: 1,2,3,4,9,8,27.

Plutarch arranged Plato's series to form a triangle and placed three additional numbers inside, which represent the sum of the numbers opposite (see below).

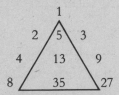

Numbers that express an astronomical fact rightfully hold a particularly high place of honor, as may be seen by Plutarch's own words in reference to his triangle:

> Now the final number of the series, which is 27, has this peculiarity, that it is equal to the sum of the preceding numbers $(1 + 2 + 3 + 4 + 9 + 8)$; it also represents the periodical number of days in which the moon completes her monthly course; the Pythagoreans have made it the tone of all their harmonic intervals.

Plutarch's statement sufficiently establishes the supreme importance of the number 27.

For most of the foregoing historical information, I am indebted to Mr. C. A. Browne, Jr., who, using his own ingenious method, created the fantastic 27 $\times$ 27 magic square shown on the following page.

352	381	326	439	468	413	274	303	248	613	642	587	700	729	674	535	564	509	118	147	92	205	234	179	40	69	14
327	353	379	414	440	466	249	275	301	588	614	640	675	701	727	510	536	562	93	119	145	180	206	232	15	41	67
380	325	354	467	412	441	302	247	276	641	586	615	728	673	702	563	508	537	146	91	120	233	178	207	68	13	42
277	306	251	355	384	329	433	462	407	538	567	512	616	645	590	694	723	668	43	72	17	121	150	95	199	228	173
252	278	304	330	356	382	408	434	460	513	539	565	591	617	643	669	695	721	18	44	70	96	122	148	174	200	226
305	250	279	383	328	357	461	406	435	566	511	540	644	589	618	722	667	696	71	16	45	149	94	123	227	172	201
436	465	410	271	300	245	358	387	332	697	726	671	532	561	506	619	648	593	202	231	176	37	66	11	124	153	98
411	437	463	246	272	298	333	359	385	672	698	724	507	533	559	594	620	646	177	203	229	12	38	64	99	125	151
464	409	438	299	244	273	386	331	360	725	670	699	560	505	534	647	592	621	230	175	204	65	10	39	152	97	126
127	156	101	214	243	188	49	78	23	361	390	335	448	477	422	283	312	257	595	624	569	682	711	656	517	546	491
102	128	154	189	215	241	24	50	76	336	362	388	423	449	475	258	284	310	570	596	622	657	683	709	492	518	544
155	100	129	242	187	216	77	22	51	389	334	363	476	421	450	311	256	285	623	568	597	710	655	684	545	490	519
52	81	26	130	159	104	208	237	182	286	315	260	364	393	338	442	471	416	520	549	494	598	627	572	676	705	650
27	53	79	105	131	157	183	209	235	261	287	313	339	365	391	417	443	469	495	521	547	573	599	625	651	677	703
80	25	54	158	103	132	236	181	210	314	259	288	392	337	366	470	415	444	548	493	522	626	571	600	704	649	678
211	240	185	46	75	20	133	162	107	445	474	419	280	309	254	367	396	341	679	708	653	514	543	488	601	630	575
186	212	238	21	47	73	108	134	160	420	446	472	255	281	307	342	368	394	654	680	706	489	515	541	576	602	628
239	184	213	74	19	48	161	106	135	473	418	447	308	253	282	395	340	369	707	652	681	542	487	516	629	574	603
604	633	578	691	720	665	526	555	500	109	138	83	196	225	170	31	60	5	370	399	344	457	486	431	292	321	266
579	605	631	666	692	718	501	527	553	84	110	136	171	197	223	6	32	58	345	371	397	432	458	484	267	293	319
632	577	606	719	664	693	554	499	528	137	82	111	224	169	198	59	4	33	398	343	372	485	430	459	320	265	294
529	558	503	607	636	581	685	714	659	34	63	8	112	141	86	190	219	164	295	324	269	373	402	347	451	480	425
504	530	556	582	608	634	660	686	712	9	35	61	87	113	139	165	191	217	270	296	322	348	374	400	426	452	478
557	502	531	635	580	609	713	658	687	62	7	36	140	85	114	218	163	192	323	268	297	401	346	375	479	424	453
688	717	662	523	552	497	610	639	584	193	222	167	28	57	2	115	144	89	454	483	428	289	318	263	376	405	350
663	689	715	498	524	550	585	611	637	168	194	220	3	29	55	90	116	142	429	455	481	264	290	316	351	377	403
716	661	690	551	496	525	638	583	612	221	166	195	56	1	30	143	88	117	482	427	456	317	262	291	404	349	378

4	9	2
3	5	7
8	1	6

Cornerstone 3 × 3

Mr. Browne originally laid out his 27 × 27 diagram in checkerboard formation with 365 white squares, or days, and 364 dark squares, or nights (total 729). We have omitted the checkerboard layout here so that the numbers may be more easily read. The number 365, the days of the year, very appropriately occupies the center of the system. (Mr. Browne's square is of great interest to Greek scholars because it throws light on an obscure passage in Plato's *Republic,* referring to a magic square, the center of which is 365.) The numbers in the rows, columns, and corner-to-corner diagonals of the central 3 × 3 square add up to 1,095, or the days of a three-year period, those of the 9 × 9 central square add up to 9,855, the days of a twenty-seven-year period—in other words, periods of years corresponding to the Pythagorean progression 1,3,9,27.

We may safely borrow the langauge of Plato and say that Mr. Browne's 27 × 27 magic square "is concerned with days and nights and months and years."

There are other notable things about this square. Not only is the 27 × 27 magic and associative, but each of the nine 9 × 9 squares, as well as the eighty-one 3 × 3 squares within the 27 × 27, is also magic as well as associative.

C. A. BROWNE JR.'S 27 X 27 COMPOSITE MAGIC SQUARE

METHOD OF CONSTRUCTION

A. Preliminary preparation:

1. Construct a 3 × 3 magic square as shown in the lower right corner of opposite page. This 3 × 3, as chosen by Mr. Browne, is the cornerstone of the progression pattern of the 729 numbers used to make up the 27 × 27 magic square. It is one of the eight possible variations of the 3 × 3 as described earlier. Any of the other 3 × 3's may be used to equal advantage.

2. Make 27 × 27 blank diagram with 729 boxes.

3. Subdivide the 27 × 27 blank diagram into nine 9 × 9 squares, and number the nine squares with the numbers 1 through 9 in the same arrangement as these numbers appear in the 3 × 3 cornerstone magic square (see opposite page).

4. Subdivide each of the 9 × 9 squares into nine 3 × 3 squares, and within each of the 3 × 3 squares, place the numbers from 1 through 9 in the same arrangement as they appear in the 3 × 3 cornerstone magic square (see opposite page).

The diagram as shown completed on the opposite page is used for reference purposes only.

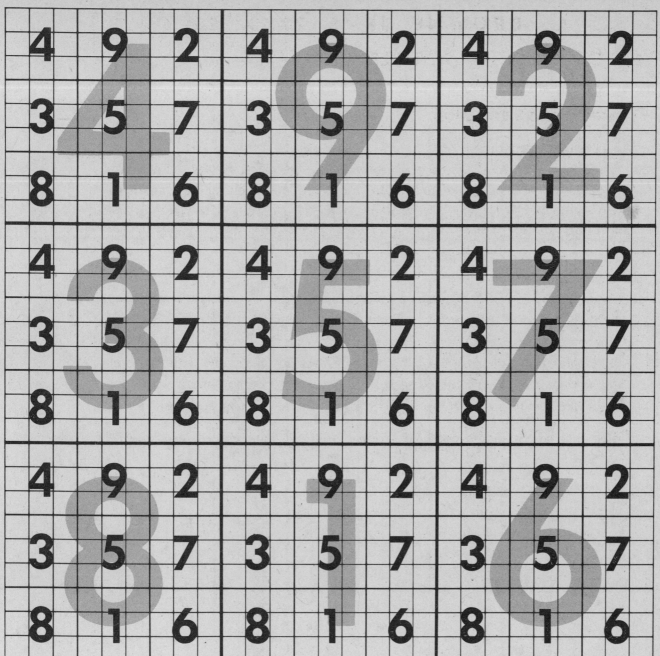

27 × 27 Reference Diagram

4	9	2
3	5	7
8	1	6

3 × 3 Cornerstone
Magic Square

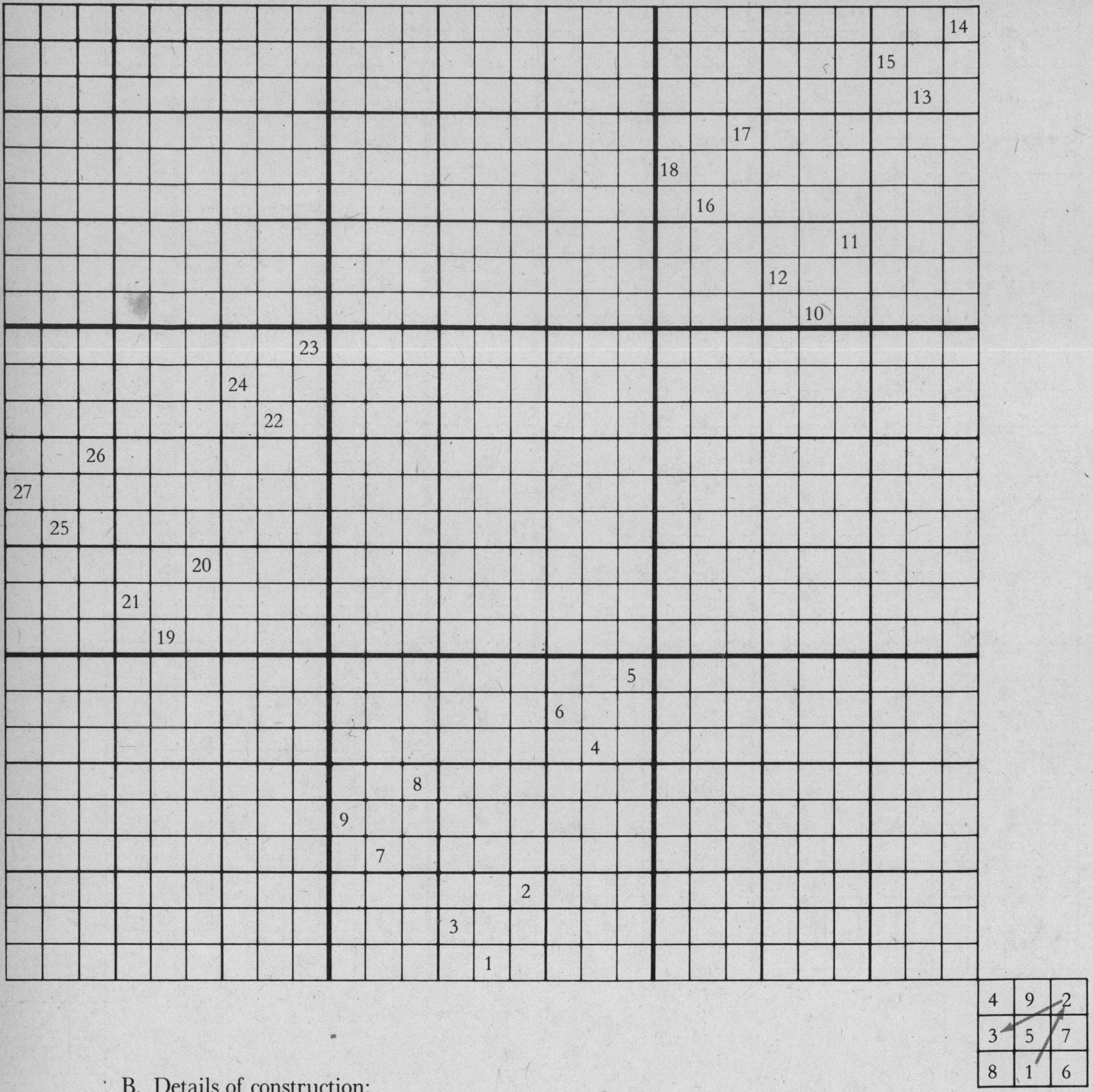

4	9	2
3	5	7
8	1	6

B. Details of construction:

1. Make a 27 × 27 blank diagram with the same subdivisions as the reference diagram but leave out the reference numbers. This diagram is used to construct the 27 × 27 magic square. The numbers 1 through 729 are placed in the 27 × 27 square in sequence groups of three (1,2,3; 4,5,6; 7,8,9; etc.).

2. The placement pattern of the first 27 numbers is shown above. (Note that this pattern follows the progression pattern indicated on the reference diagram.)

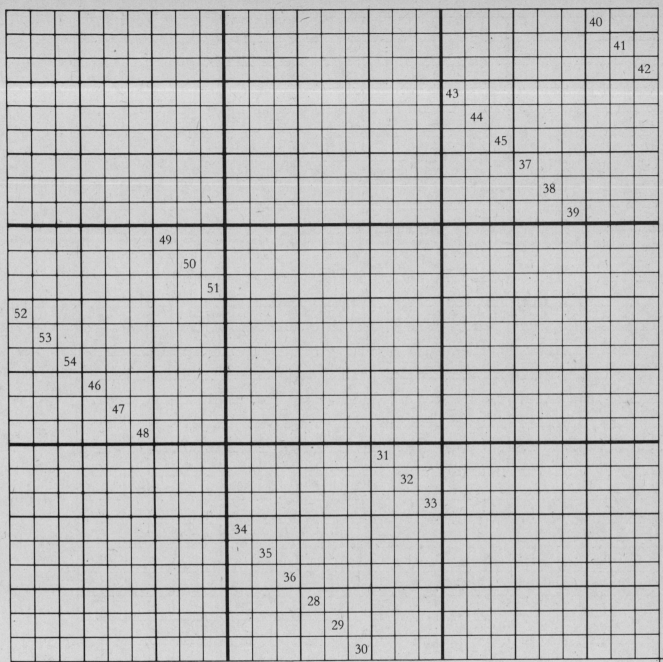

Second Group of 27 Numbers (28–54), Showing Pattern of Construction

4	9	2
3	5	7
8	1	6

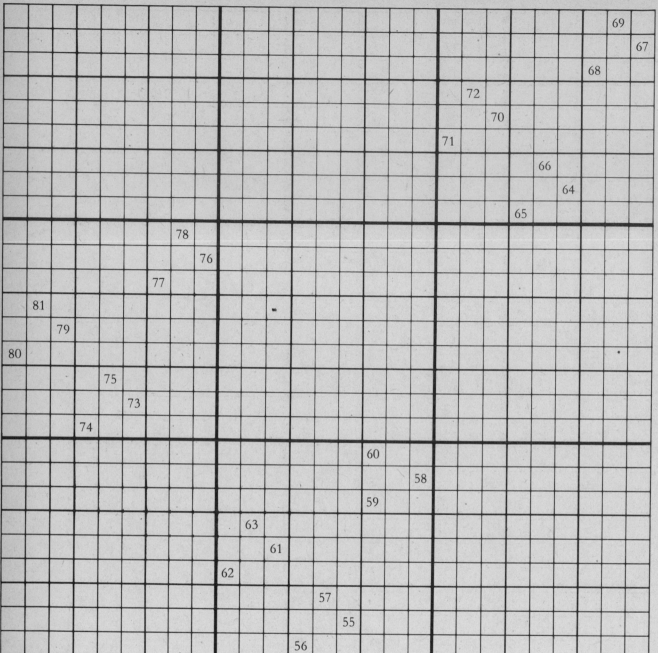

Third Group of 27 Numbers (55–81), Showing Pattern of Construction

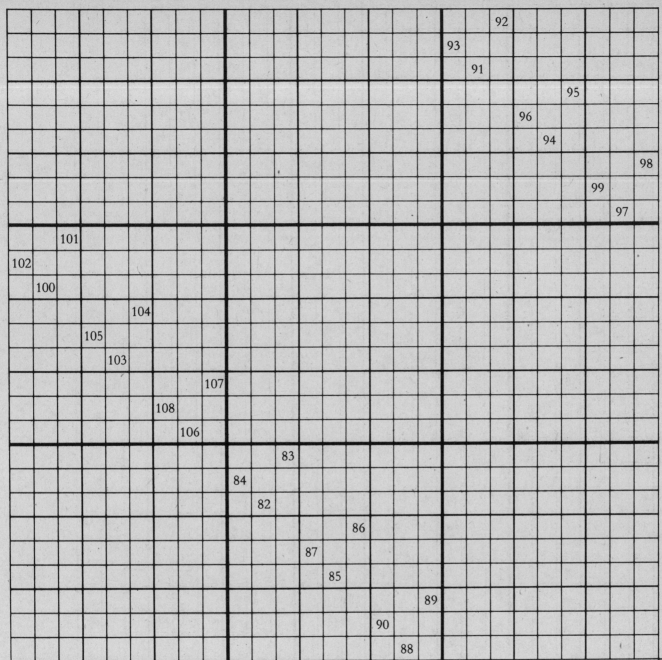

Fourth Group of 27 Numbers (82–108), Showing Pattern of Construction

4	9	2
3	5	7
8	1	6

A grid showing the placement of numbers 109-135 in a magic square construction pattern. The numbers are placed diagonally:

Top-right quadrant: 118, 119, 120, 121, 122, 123, 124, 125, 126

Left-middle quadrant: 127, 128, 129, 130, 131, 132, 133, 134, 135

Bottom-middle quadrant: 109, 110, 111, 112, 113, 114, 115, 116, 117

Fifth Group of 27 Numbers (109–135), Showing Pattern of Construction

4	9	2
3	5	7
8	1	6

Sixth Group of 27 Numbers (136–162), Showing Pattern of Construction

4	9	2
3	5	7
8	1	6

Seventh Group of 27 Numbers (163–189), Showing Pattern of Construction

4	9	2
3	5	7
8	1	6

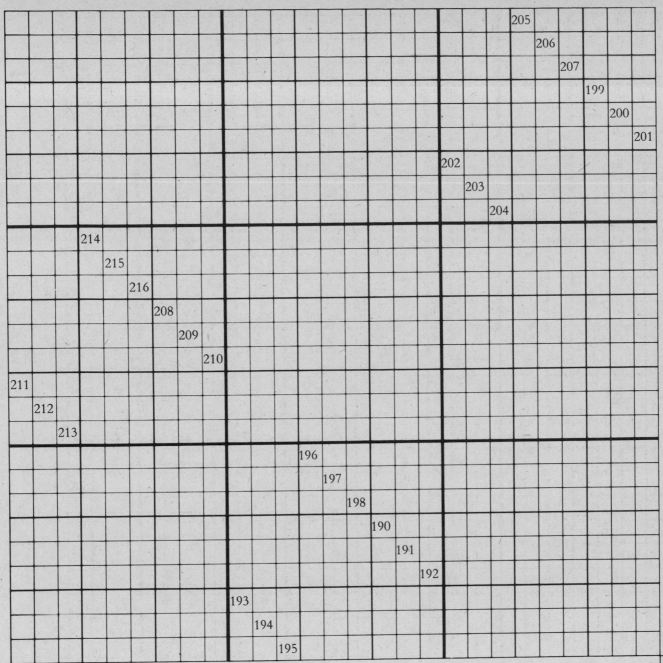

Eighth Group of 27 Numbers (190–216), Showing Pattern of Construction

4	9	2
3	5	7
8	1	6

Ninth Group of 27 Numbers (217–243), Showing Pattern of Construction

4	9	2
3	5	7
8	1	6

This completes construction of one-third of the 27 × 27 square (see following page).

ANOTHER METHOD OF CONSTRUCTING
C. A. BROWNE JR.'S 27 × 27 MAGIC SQUARE

The previously shown method of constructing Browne's 27 × 27 magic square was shown in great detail to emphasize the fact that the magic-square freak who takes the trouble to construct this square on his own will find the effort most illuminating and rewarding. However, there is a simpler but more tedious method that achieves the same result. It requires a preliminary square, which means all 729 numbers must be written twice.

1. Lay out a preliminary 27 × 27 diagram with subdivisions as before and arrange the numbers 1 through 729 in numerical order (1,2,3,4,5,6,7, and so forth), beginning with 1 in the upper left corner and ending with 729 in the lower right corner (see figure 1 on page 104).

2. Shift the nine largest squares (9 × 9) into positions indicated by 3 × 3 cornerstone.

3. Repeat the process with the subdivisions of the 9 × 9 squares and so on down to complete the 27 × 27 magic square (see figures 2, 3, and 4).

Figure 1: The top three 9 × 9 squares of preliminary 27 × 27 diagram, showing placement of numbers in numerical order. The first 9 × 9 square has been completed, the second and third indicated.

Figure 2: The bottom row of three 9 × 9 squares of 27 × 27 diagram, showing how the first 9 × 9 square from the diagram above has been shifted to the middle of the bottom row to occupy position number 1 as indicated by 3 × 3 cornerstone square.

Figure 3: The nine 3 × 3 squares of figure 2, showing how positions of each have been shifted to occupy positions as indicated by 3 × 3 cornerstone square.

Figure 4: The nine 3 × 3 squares of figure 3, showing how numbers in each square have been shifted to occupy positions as indicated by 3 × 3 cornerstone square.

Figures 1, 2, 3, and 4 show only the process of getting the numbers of the first 9 × 9 square in magic-square order for the 27 × 27. To proceed, shift number two 9 × 9 square to number two position in the 27 × 27 as indicated by the 3 × 3 cornerstone square, and so forth.

Top-right 9×9 block

118	147	92	205	234	179	40	69	14
93	119	145	180	206	232	15	41	67
146	91	120	233	178	207	68	13	42
43	72	17	121	150	95	199	228	173
18	44	70	96	122	148	174	200	226
71	16	45	149	94	123	227	172	201
202	231	176	37	66	11	124	153	98
177	203	229	12	38	64	99	125	151
230	175	204	65	10	39	152	97	126

Middle-left 9×9 block

127	156	101	214	243	188	49	78	23
102	128	154	189	215	241	24	50	76
155	100	129	242	187	216	77	22	51
52	81	26	130	159	104	208	237	182
27	53	79	105	131	157	183	209	235
80	25	54	158	103	132	236	181	210
211	240	185	46	75	20	133	162	107
186	212	238	21	47	73	108	134	160
239	184	213	74	19	48	161	106	135

Bottom-middle 9×9 block

109	138	83	196	225	170	31	60	5
84	110	136	171	197	223	6	32	58
137	82	111	224	169	198	59	4	33
34	63	8	112	141	86	190	219	164
9	35	61	87	113	139	165	191	217
62	7	36	140	85	114	218	163	192
193	222	167	28	57	2	115	144	89
168	194	220	3	29	55	90	116	142
221	166	195	56	1	30	143	88	117

The 27 × 27 Magic Square One-Third Completed

4	9	2
3	5	7
8	1	6

Begin next pattern by placing 244 directly above 243 (the last number placed) and proceed to complete the 27 × 27 magic square according to pattern indicated by reference diagram.

3 × 3 Cornerstone Magic Square

4	9	2
3	5	7
8	1	6

Top Row of 27 × 27 Preliminary Square

1	2	3	4	5	6	7	8	9	10	11	12	13	14	15	16	17	18	19	20	21	22	23	24	25	26	27
28	29	30	31	32	33	34	35	36	37	Etc	→															54
55	56	57	58	59	60	61	62	63	64	→																81
82	83	84	85	86	87	88	89	90	91	→																108
109	110	111	112	113	114	115	116	117	118	→																135
136	137	138	139	140	141	142	143	144	145	→																162
163	164	165	166	167	168	169	170	171	172	→																189
190	191	192	193	194	195	196	197	198	199	→																216
217	218	219	220	221	222	223	224	225	226	→																243

Figure 1

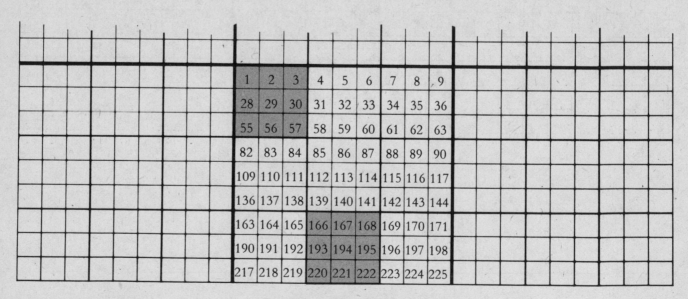

Bottom Row of 27 × 27 Square

Figure 2

82	83	84	169	170	171	4	5	6
109	110	111	196	197	198	31	32	33
136	137	138	223	224	225	58	59	60
7	8	9	85	86	87	163	164	165
34	35	36	112	113	114	190	191	192
61	62	63	139	140	141	217	218	219
166	167	168	1	2	3	88	89	90
193	194	195	28	29	30	115	116	117
220	221	222	55	56	57	142	143	144

Figure 3

109	138	83	196	225	170	31	60	5
84	110	136	171	197	223	6	32	58
137	82	111	224	169	198	59	4	33
34	63	8	112	141	86	190	219	164
9	35	61	87	113	139	165	191	217
62	7	36	140	85	114	218	163	192
193	222	167	28	57	2	115	144	89
168	194	220	3	29	55	90	116	142
221	166	195	56	1	30	143	88	117

Figure 4

PART THREE

THE CONSTRUCTION OF EVEN-ORDER MAGIC SQUARES

The construction of even-order magic squares is a brand-new and completely different ball game—more mysterious, more fascinating, and more challenging.

Whereas nearly all the methods we have shown for odd-order magic squares may be used to construct odd-order squares of any size, this is generally not true of even-order squares.

The primary reason for this seems to lie in the fact that even-order squares have less in common with each other than odd-order squares. For example, the 8 × 8 square: the number 8 can be divided by both 2 and 4. The 6 × 6 square: 6 can be divided by 2 but not by 4. Squares based on 6, 10, 14, and so forth, which can be divided by 2 but not by 4, conceal particularly complex building secrets not readily evident to the casual observer. Also, even-order squares that can be divided by both odd and even numbers (for example, 6 × 6, 10 × 10, 12 × 12) have their own peculiarities.

One simple way to observe the basic differences between odd- and even-order squares is to arrange the numbers in each in numerical order (1,2,3,4,5, etc.) and compare them (see below).

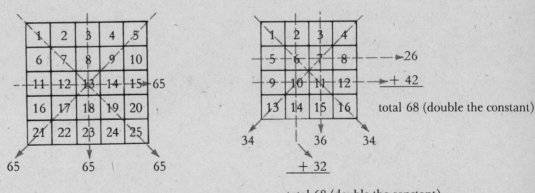

5 × 5

4 × 4

In the odd-order square above (5 × 5), it will be seen that the sums of the numbers in the two corner-to-corner diagonals as well as the middle column and the middle row all total the constant 65. Also the sum of any pair of diametrically opposed numbers equidistant from the center is 26 (the sum of the first and last numbers of the series).

Similarly, in the even-order square (4 × 4), it will be seen that the sums of the numbers in the diagonals total the constant 34 and that the sums of the paired numbers equidistant from the center total 17 (the sum of the first and last numbers of the series). Here, for our purposes, the similarity ends. But it is interesting to note that the sum of the numbers in the two middle columns is 68—exactly double the amount of the constant 34. This is also true of the two middle rows. This information may prove useful in understanding the construction of even-order squares.

Method A for 4 x 4

Preliminary square A:

1. Fill in the boxes of 4 × 4 diagram with the numbers 1 through 16 in their natural sequence (1,2,3,4,5, etc.) (see below).

Magic square B:

1. The magic square B is the same as preliminary square A except that the order of the numbers in the two corner-to-corner diagonals has been reversed (see below).

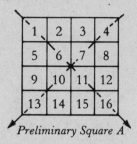

Preliminary Square A

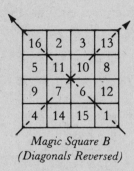

Magic Square B
(Diagonals Reversed)

Method B for 4 × 4

Preliminary square A:

1. Fill in the boxes of the 4 × 4 diagram with the numbers 1 through 16 in their natural sequence (1,2,3,4,5, etc.).

Magic square B:

1. The numbers in the corner-to-corner diagonals remain in the same arrangement as in preliminary square A.
2. Reverse the order of the numbers in each pair of remaining complementary numbers (2 and 15, 8 and 9, etc.).

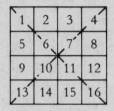

Preliminary Square A

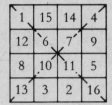

Magic Square B

Having made two apparently different 4 × 4 magic squares by 4 × 4 methods A and B, it may come as a surprise and perhaps a disappointment to discover that the second square made by method B is merely a disguised version of the original. This becomes evident upon making sequence designs of the two squares (see below).

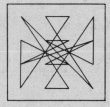

16	2	3	13
5	11	10	8
9	7	6	12
4	14	15	1

Original

Magic square made by method A

Sequence Designs
Identical

1	15	14	4
12	6	7	9
8	10	11	5
13	3	2	16

Disguised

Magic square made by method B

METHOD C FOR 4 X 4:
VARIATIONS OF INTERCHANGE

AS EXPLAINED BY MAURICE KRAITCHIK

1. Interchange border rows of an original magic square.
2. Interchange border columns of an original magic square.
3. Interchange border columns of original magic square and *then* interchange border rows of the newly made square.
4. Interchange quadrants of an original magic square.

Following are examples of each.

Original

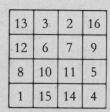

Top and Bottom
Rows Interchanged

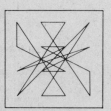

Sequence Design

Step 1

1	15	14	4
12	6	7	9
8	10	11	5
13	3	2	16

Original

4	15	14	1
9	6	7	12
5	10	11	8
16	3	2	13

Left and Right
Columns Interchanged

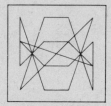

Sequence Design

Step 2

4	15	14	1
9	6	7	12
5	10	11	8
16	3	2	13

Left and Right Columns
Interchanged; Then

16	3	2	13
9	6	7	12
5	10	11	8
4	15	14	1

Top and Bottom
Rows Interchanged

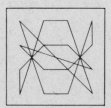

Sequence Design

Step 3

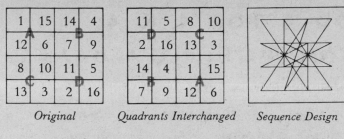

Original	Quadrants Interchanged	Sequence Design

Step 4

It will be seen by comparing the sequence designs that each of the four variations of the interchange produced an original magic square.

Project: Try to discover additional variations of the interchange.

Method D for 4 x 4:
ROTATION WITHIN QUADRANTS

INVENTED (OR REINVENTED) BY HOWES AND MORAN

This method uses one original magic square from which other magic squares are derived. For present purposes, we find it appropriate to refer to the original magic square as a mother (no slight intended) and the derived squares as off-spring. The original mother was constructed by method B.

1. Using heavier lines, divide the mother into four quadrants and number the quadrants 1,2,3,4 clockwise. Place arrows pointing clockwise between each of the four numbers in each quadrant.

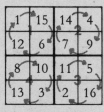

Mother

2. Referring to the mother above, follow the arrows in quadrant 1 and make the top row of offspring square as shown below. Start with the 1 in upper left corner.

1	15	6	12

Offspring Top Row Completed

3. Fill in the second row of the offspring square with the numbers in quadrant 2. Again start with the number in the upper left corner of quadrant and continue clockwise as before (see below).

1	15	6	12
14	4	9	7

Offspring Second Row Completed

4. Following the same pattern, fill in the third and fourth rows with the numbers within quadrants three and four. Offspring 1 is now complete as shown.

1	15	6	12
14	4	9	7
11	5	16	2
8	10	3	13

Offspring 1

5. To construct offspring 2, refer again to the above mother, but instead of starting with 1, begin with the 15 (box 2) in quadrant 1, and proceeding clockwise as before, fill in row one with the numbers within this quadrant.

15	6	12	1

6. Complete offspring 2 by filling in the remaining rows using the same system.

15	6	12	1
4	9	7	14
5	16	2	11
10	3	13	8

Offspring 2

7. To construct offspring 3, use same pattern as above but start with box 3 of each successive quadrant and proceed clockwise.

6	12	1	15
9	7	14	4
16	2	11	5
3	13	8	10

Offspring 3

8. Offspring 4 is made by starting with box 4 of quadrant one, etc.

12	1	15	6
7	14	4	9
2	11	5	16
13	8	10	3

Offspring 4

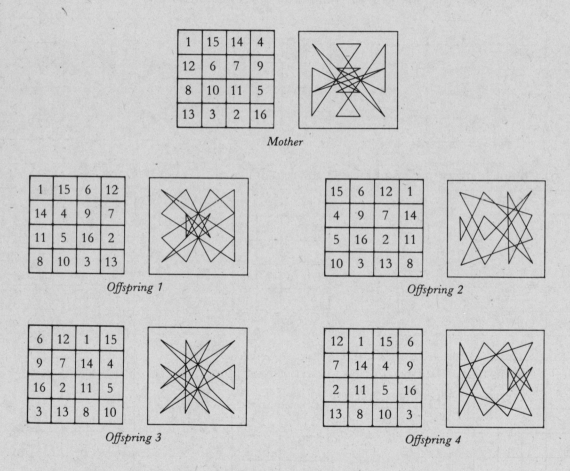

1	15	14	4
12	6	7	9
8	10	11	5
13	3	2	16

Mother

1	15	6	12
14	4	9	7
11	5	16	2
8	10	3	13

Offspring 1

15	6	12	1
4	9	7	14
5	16	2	11
10	3	13	8

Offspring 2

6	12	1	15
9	7	14	4
16	2	11	5
3	13	8	10

Offspring 3

12	1	15	6
7	14	4	9
2	11	5	16
13	8	10	3

Offspring 4

The mother above was constructed by method B. Her four offspring were shown on previous pages but without the identifying sequence design.

Magic-Square Families

Magic squares may be classified by families and family differences made visible by drawing family designs for easy identification. Family designs are made by drawing a straight line between the complementary numbers of that square. Example: The lowest and highest numbers in the 4 × 4 square are 1 and 16 (1 + 16 = 17). By drawing a straight line between each pair of numbers totaling 17, we are able to draw the family design for any particular square.

As shown in the previous method C for 4 × 4 squares, we can induce certain mother squares to give birth to offspring by revolving the numbers in their quadrants. Accepting this premise leads us into some strange and fascinating analogies between magic squares and certain aspects of genetics, such as breeding, cell division, and so forth.

4 × 4 magic squares may be divided into twelve families, as shown on the following two pages. (No one as yet has been able to discover additional families.) For purposes of identification, we have chosen family names from the names of Greek goddesses and gods.

From a genetics standpoint, it is interesting to note that the first six families are purebred and fertile—purebred because each family design is unique and fertile because any member of any one of the six families is capable of producing offspring by our method of breeding.

The second six families are hybrid and sterile—hybrid because their family design shows them to be of mixed breed. This may be seen by observing the Cottus family design, which is obviously some sort of a mixture of the Elara and Asteria families (see family designs). The same is true of the rest of the second six families. They are sterile because they are not capable of producing offspring by our method of breeding.

CLASSIFICATION OF 4 X 4 MAGIC SQUARES
BY FAMILIES

A straight line joins each pair of complementary numbers. There are twelve 4 × 4 families. Six are purebred and fertile. Six are hybrid and sterile.

16	1	13	4
7	10	6	11
2	15	3	14
9	8	12	5

Elara

1	7	14	12
10	16	5	3
15	9	4	6
8	2	11	13

Asteria

4	1	13	16
14	15	3	2
11	10	6	7
5	8	12	9

Hestia

1	8	10	15
14	11	5	4
7	2	16	9
12	13	3	6

Hera

1	13	4	16
8	12	5	9
14	2	15	3
11	7	10	6

Demeter

1	15	14	4
12	6	7	9
8	10	11	5
13	3	2	16

Niobe

The six squares above are purebred and fertile.

16	1	12	5
2	11	6	15
7	14	3	10
9	8	13	4

Thaumas

12	4	13	5
1	9	16	8
15	7	2	10
6	14	3	11

Nemesis

11	14	3	6
8	9	16	1
10	7	2	15
5	4	13	12

Arges

1	2	16	15
13	14	4	3
12	7	9	6
8	11	5	10

Eris

5	1	12	16
10	14	3	7
15	11	6	2
4	8	13	9

Moros

2	15	1	16
11	10	8	5
14	3	13	4
7	6	12	9

Cottus

The six squares above are hybrid (compare with first six) and sterile.

Magic-Square Genetics

Some initial steps toward an exploration of the concept of magic-square genetics are shown by the following charts. An interesting project may lie ahead for the curious magic-square freak who wishes to delve deeper into the subject and perhaps carry it as far as the twelfth generation. As may be seen by our limited investigation, each succeeding generation is different from its forebears but still retains some similarities.

Queen Mother Anne of the Niobe family represents he first generation. She was created by 4 × 4 method B. All her descendants are bred by 4 × 4 method C (see chart A).

Some observations regarding the second generation (see chart A):

1. Anne has eight daughters, including two sets of R.R. twins (reflections-rotations).
2. All eight daughters are of the Hera family.
3. The names of all eight daughters begin with the letter B, which signifies that they are of the second generation. (The letter C signifies the third generation, etc.)

Some observations regarding the third generation (see Charts A1–A8):

1. Each of Anne's eight daughters (second generation) bears four daughters (third generation), including one set of R.R. twins.
2. All thirty-two members of third generation are of the Hestia family.
3. Comparison of Charts A1 and A3 reveals that similarly placed daughters in each are R.R. "cousins."
4. The above is true of Charts A2–A4, A5–A7, and A6–A8.
5. This is *not* true of Charts A3–A5, A4–A6.

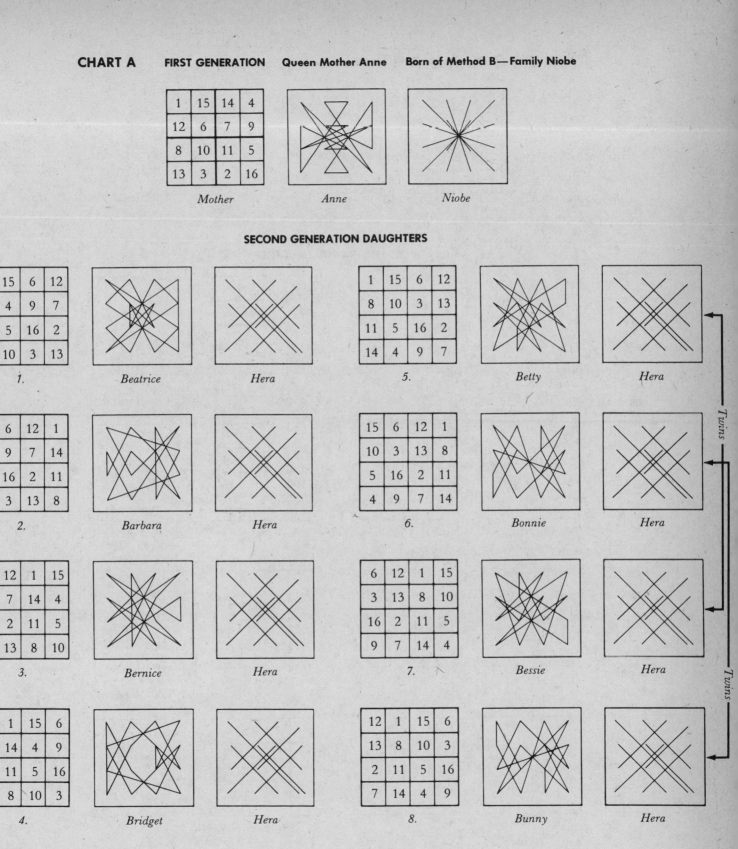

Mother   Anne   Niobe

SECOND GENERATION DAUGHTERS

1.   Beatrice   Hera   5.   Betty   Hera

2.   Barbara   Hera   6.   Bonnie   Hera

3.   Bernice   Hera   7.   Bessie   Hera

4.   Bridget   Hera   8.   Bunny   Hera

Twins

Twins

Anne of the Niobe family produces eight daughters, all of the Hera family.

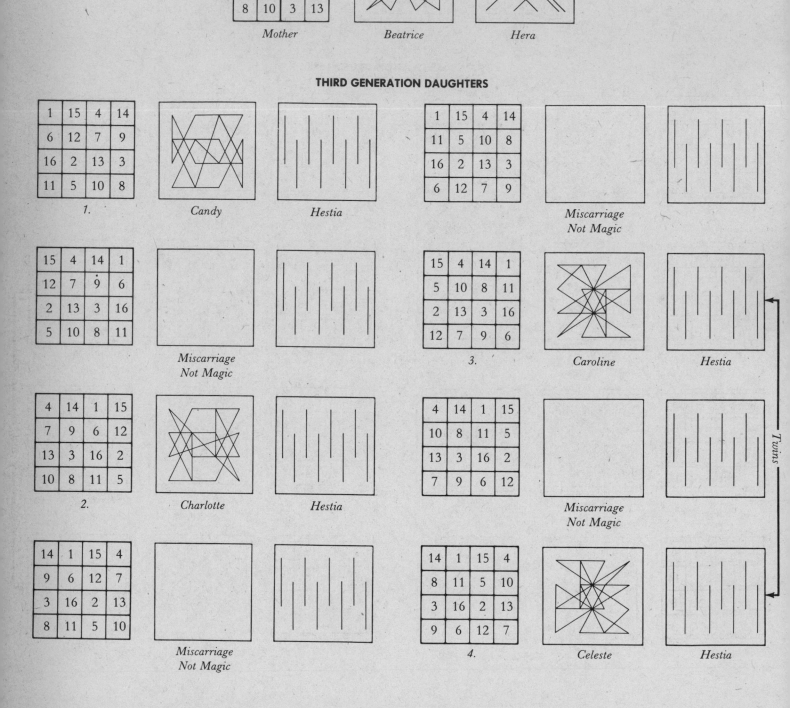

1	15	6	12
14	4	9	7
11	5	16	2
8	10	3	13

Mother   *Beatrice*   *Hera*

THIRD GENERATION DAUGHTERS

1	15	4	14
6	12	7	9
16	2	13	3
11	5	10	8

1.   *Candy*   *Hestia*

1	15	4	14
11	5	10	8
16	2	13	3
6	12	7	9

Miscarriage Not Magic

15	4	14	1
12	7	9	6
2	13	3	16
5	10	8	11

Miscarriage Not Magic   *Hestia*

15	4	14	1
5	10	8	11
2	13	3	16
12	7	9	6

3.   *Caroline*   *Hestia*

4	14	1	15
7	9	6	12
13	3	16	2
10	8	11	5

2.   *Charlotte*   *Hestia*

4	14	1	15
10	8	11	5
13	3	16	2
7	9	6	12

Miscarriage Not Magic   *Hestia*

14	1	15	4
9	6	12	7
3	16	2	13
8	11	5	10

Miscarriage Not Magic

14	1	15	4
8	11	5	10
3	16	2	13
9	6	12	7

4.   *Celeste*   *Hestia*

Twins

Beatrice of the Hera family has four daughters, all of the Hestia family.

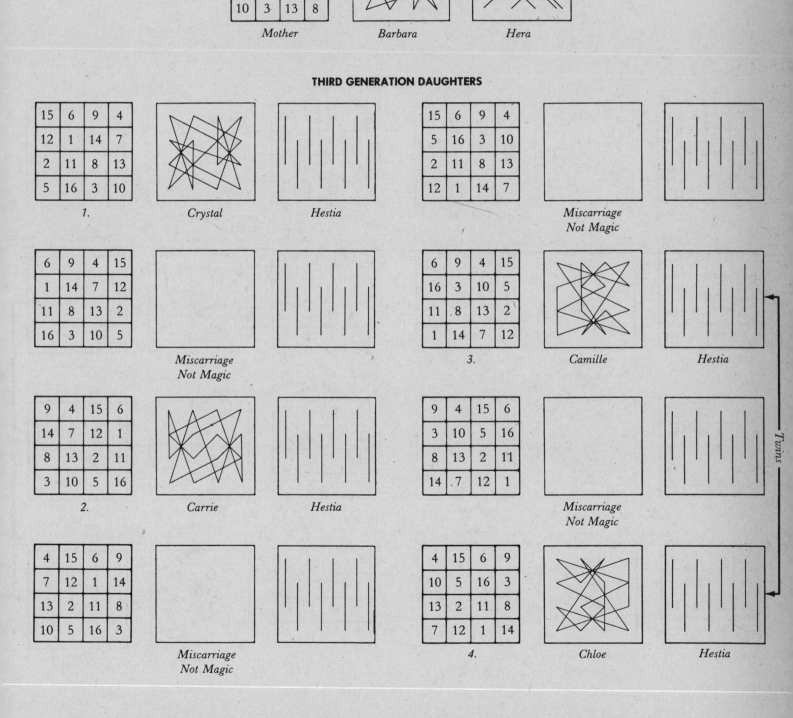

15	6	12	1
4	9	7	14
5	16	2	11
10	3	13	8

Mother   *Barbara*   *Hera*

THIRD GENERATION DAUGHTERS

15	6	9	4
12	1	14	7
2	11	8	13
5	16	3	10

1.   *Crystal*   *Hestia*

15	6	9	4
5	16	3	10
2	11	8	13
12	1	14	7

Miscarriage Not Magic

6	9	4	15
1	14	7	12
11	8	13	2
16	3	10	5

Miscarriage Not Magic

6	9	4	15
16	3	10	5
11	8	13	2
1	14	7	12

3.   *Camille*   *Hestia*

9	4	15	6
14	7	12	1
8	13	2	11
3	10	5	16

2.   *Carrie*   *Hestia*

9	4	15	6
3	10	5	16
8	13	2	11
14	7	12	1

Miscarriage Not Magic

4	15	6	9
7	12	1	14
13	2	11	8
10	5	16	3

Miscarriage Not Magic

4	15	6	9
10	5	16	3
13	2	11	8
7	12	1	14

4.   *Chloe*   *Hestia*

Twins

Barbara of the Hera family has four daughters, all of the Hestia family.

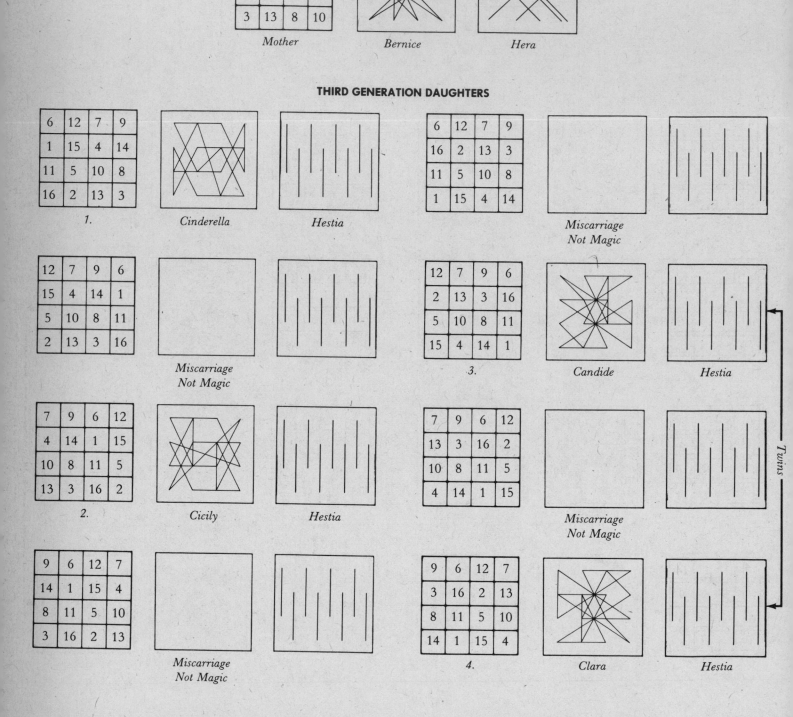

6	12	1	15
9	7	14	4
16	2	11	5
3	13	8	10

Mother

Bernice

Hera

THIRD GENERATION DAUGHTERS

6	12	7	9
1	15	4	14
11	5	10	8
16	2	13	3

1.

Cinderella

Hestia

6	12	7	9
16	2	13	3
11	5	10	8
1	15	4	14

Miscarriage
Not Magic

12	7	9	6
15	4	14	1
5	10	8	11
2	13	3	16

Miscarriage
Not Magic

12	7	9	6
2	13	3	16
5	10	8	11
15	4	14	1

3.

Candide

Hestia

7	9	6	12
4	14	1	15
10	8	11	5
13	3	16	2

2.

Cicily

Hestia

7	9	6	12
13	3	16	2
10	8	11	5
4	14	1	15

Miscarriage
Not Magic

9	6	12	7
14	1	15	4
8	11	5	10
3	16	2	13

Miscarriage
Not Magic

9	6	12	7
3	16	2	13
8	11	5	10
14	1	15	4

4.

Clara

Hestia

Twins

Bernice of the Hera family has four daughters, all of the Hestia family.

SECOND GENERATION Bridget, Fourth Daughter of Anne

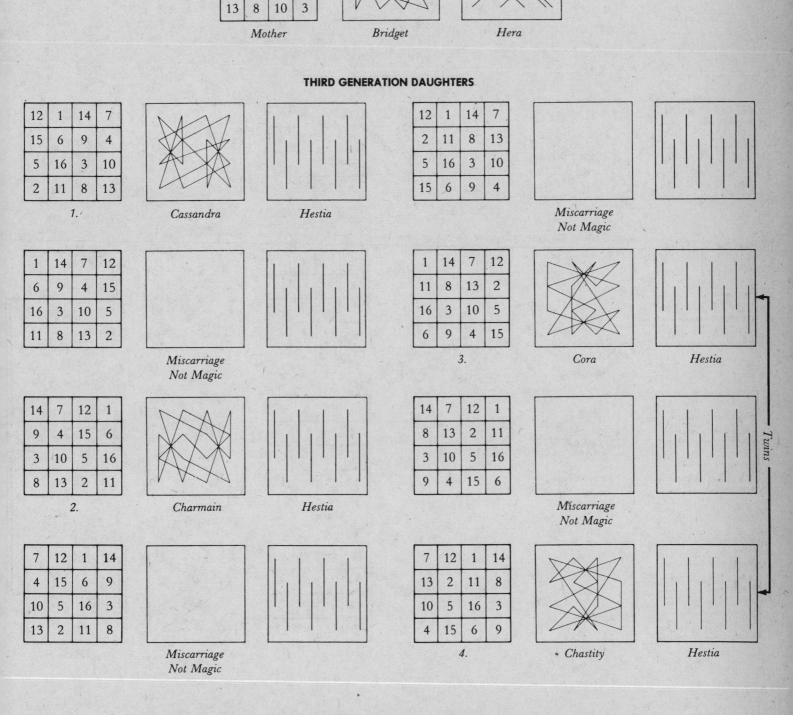

12	1	15	6
7	14	4	9
2	11	5	16
13	8	10	3

Mother

Bridget

Hera

THIRD GENERATION DAUGHTERS

12	1	14	7
15	6	9	4
5	16	3	10
2	11	8	13

1.

Cassandra

Hestia

12	1	14	7
2	11	8	13
5	16	3	10
15	6	9	4

Miscarriage Not Magic

1	14	7	12
6	9	4	15
16	3	10	5
11	8	13	2

Miscarriage Not Magic

1	14	7	12
11	8	13	2
16	3	10	5
6	9	4	15

3.

Cora

Hestia

14	7	12	1
9	4	15	6
3	10	5	16
8	13	2	11

2.

Charmain

Hestia

14	7	12	1
8	13	2	11
3	10	5	16
9	4	15	6

Miscarriage Not Magic

7	12	1	14
4	15	6	9
10	5	16	3
13	2	11	8

Miscarriage Not Magic

7	12	1	14
13	2	11	8
10	5	16	3
4	15	6	9

4.

Chastity

Hestia

Twins

Bridget of the Hera family has four daughters, all of the Hestia family.

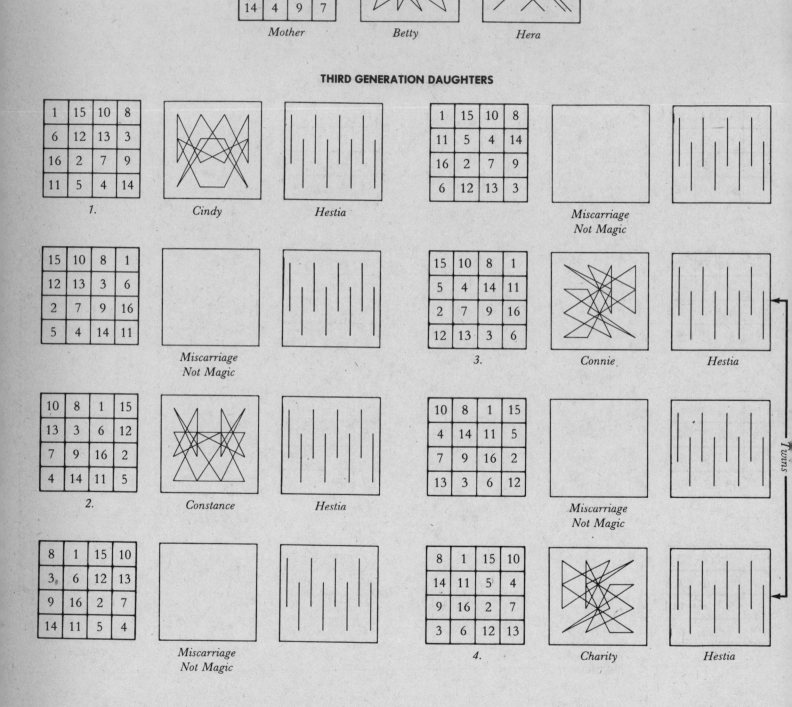

1	15	6	12
8	10	3	13
11	5	16	2
14	4	9	7

Mother

Betty

Hera

THIRD GENERATION DAUGHTERS

1	15	10	8
6	12	13	3
16	2	7	9
11	5	4	14

1.

Cindy

Hestia

1	15	10	8
11	5	4	14
16	2	7	9
6	12	13	3

Miscarriage Not Magic

15	10	8	1
12	13	3	6
2	7	9	16
5	4	14	11

Miscarriage Not Magic

15	10	8	1
5	4	14	11
2	7	9	16
12	13	3	6

3.

Connie

Hestia

10	8	1	15
13	3	6	12
7	9	16	2
4	14	11	5

2.

Constance

Hestia

10	8	1	15
4	14	11	5
7	9	16	2
13	3	6	12

Miscarriage Not Magic

8	1	15	10
3	6	12	13
9	16	2	7
14	11	5	4

Miscarriage Not Magic

8	1	15	10
14	11	5	4
9	16	2	7
3	6	12	13

4.

Charity

Hestia

Twins

Betty of the Hera family has four daughters, all of the Hestia family.

SECOND GENERATION   Bonnie, Sixth Daughter of Anne

15	6	12	1
10	3	13	8
5	16	2	11
4	9	7	14

Mother

Bonnie

Hera

THIRD GENERATION DAUGHTERS

15	6	3	10
12	1	8	13
2	11	14	7
5	16	9	4

1.

Charlene

Hestia

15	6	3	10
5	16	9	4
2	11	14	7
12	1	8	3

Miscarriage
Not Magic

Hestia

6	3	10	15
1	8	13	12
11	14	7	2
16	9	4	5

Miscarriage
Not Magic

6	3	10	15
16	9	4	5
11	14	7	2
1	8	13	12

3.

Carol

Hestia

3	10	15	6
8	13	12	1
14	7	2	11
9	4	5	16

2.

Cerise

Hestia

3	10	15	6
9	4	5	16
14	7	2	11
8	13	12	1

Miscarriage
Not Magic

10	15	6	3
13	12	1	8
7	2	11	14
4	5	16	9

Miscarriage
Not Magic

10	15	6	3
4	5	16	9
7	2	11	14
13	12	1	8

4.

Cherry

Hestia

Twins

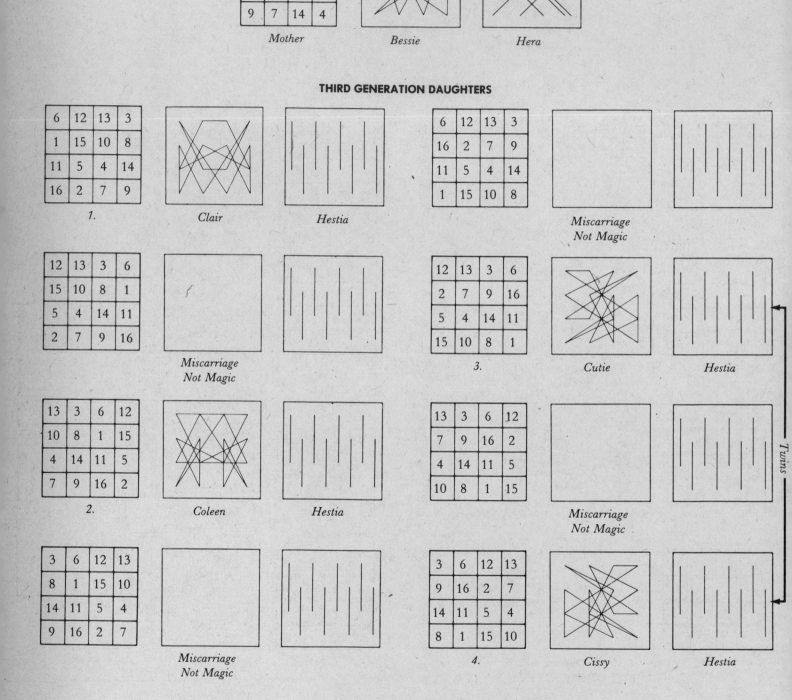

Mother

6	12	1	15
3	13	8	10
16	2	11	5
9	7	14	4

Bessie

Hera

THIRD GENERATION DAUGHTERS

1.

6	12	13	3
1	15	10	8
11	5	4	14
16	2	7	9

Clair

Hestia

6	12	13	3
16	2	7	9
11	5	4	14
1	15	10	8

Miscarriage
Not Magic

2.

12	13	3	6
15	10	8	1
5	4	14	11
2	7	9	16

Miscarriage
Not Magic

3.

12	13	3	6
2	7	9	16
5	4	14	11
15	10	8	1

Cutie

Hestia

13	3	6	12
10	8	1	15
4	14	11	5
7	9	16	2

Coleen

Hestia

13	3	6	12
7	9	16	2
4	14	11	5
10	8	1	15

Miscarriage
Not Magic

4.

3	6	12	13
8	1	15	10
14	11	5	4
9	16	2	7

Miscarriage
Not Magic

3	6	12	13
9	16	2	7
14	11	5	4
8	1	15	10

Cissy

Hestia

Twins

Bessie of the Hera family has four daughters, all of the Hestia family.

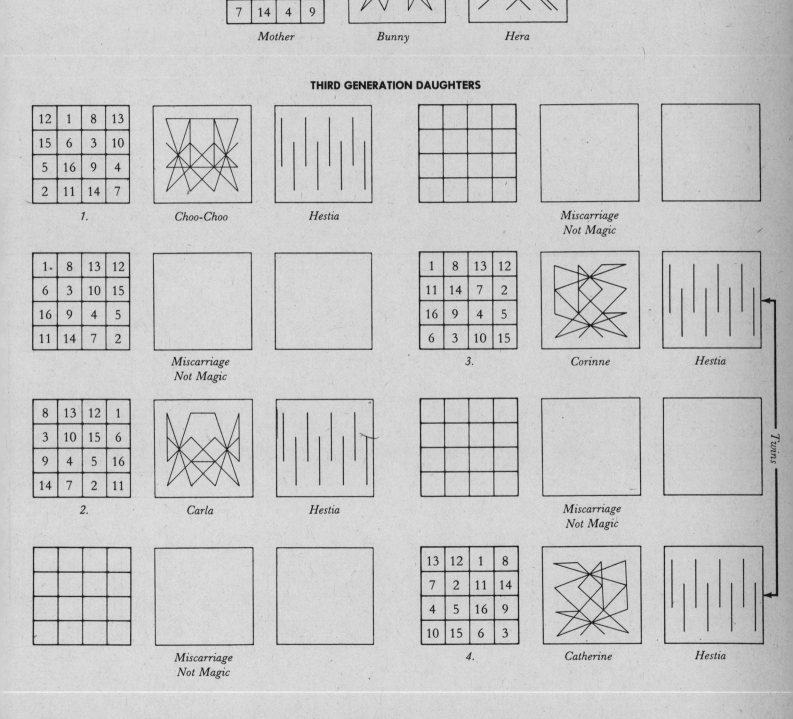

12	1	15	6
13	8	10	3
2	11	5	16
7	14	4	9

Mother

Bunny

Hera

THIRD GENERATION DAUGHTERS

12	1	8	13
15	6	3	10
5	16	9	4
2	11	14	7

1.

Choo-Choo

Hestia

Miscarriage Not Magic

1	8	13	12
6	3	10	15
16	9	4	5
11	14	7	2

Miscarriage Not Magic

1	8	13	12
11	14	7	2
16	9	4	5
6	3	10	15

3.

Corinne

Hestia

8	13	12	1
3	10	15	6
9	4	5	16
14	7	2	11

2.

Carla

Hestia

Miscarriage Not Magic

Miscarriage Not Magic

13	12	1	8
7	2	11	14
4	5	16	9
10	15	6	3

4.

Catherine

Hestia

Twins

Bunny of the Hera family has four daughters, all of the Hestia family.

Some observations regarding the fourth generation (see Charts B1–B4):

1. Charts B1–B4 cover only one branch (or one-eighth) of the fourth generation.
2. The fourth generation is divided evenly between *two* families—Demeter and Elara.
3. Each group of four daughters on Charts B1 and B2 includes one set of R.R. twins.
4. Each group of four daughters on Charts B2 and B3 includes *two* sets of R.R. twins.
5. Daughter 1 on Chart B1 is an identical cousin of daughter 2 on Chart B2.
6. Daughter 2 on Chart B1 is an identical cousin of daughter 1 on Chart B2.
7. Daughter 3 on Chart B1 is an identical cousin of daughter 4 on Chart B2.
8. Daughter 4 on Chart B1 is an identical cousin of daughter 3 on Chart B2.
9. Daughter 3 on Chart B3 is an R.R. cousin of daughter 3 on Chart B4.
10. Daughter 4 on Chart B3 is an R.R. cousin of daughter 4 on Chart B4.

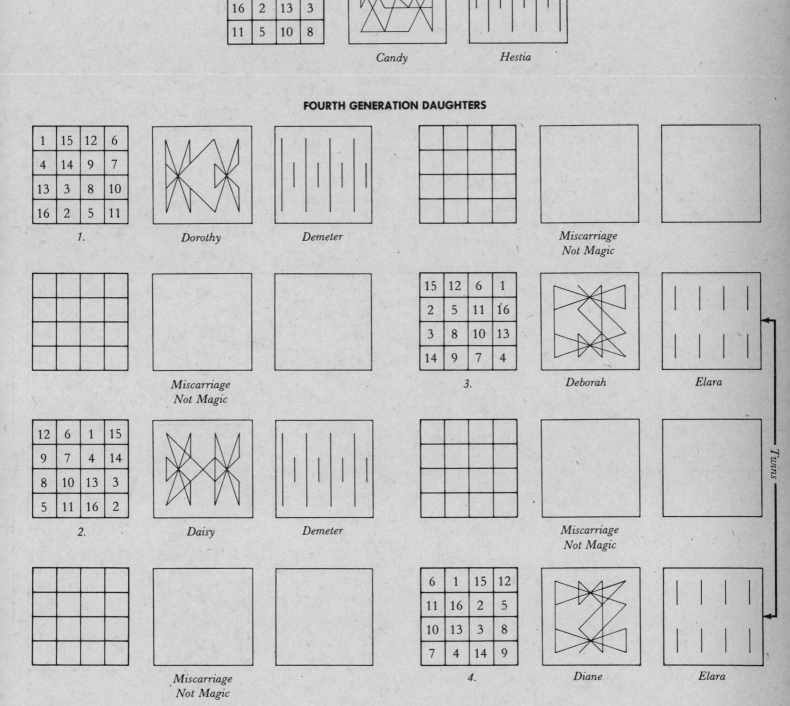

1	15	4	14
6	12	7	9
16	2	13	3
11	5	10	8

Candy   *Hestia*

FOURTH GENERATION DAUGHTERS

1	15	12	6
4	14	9	7
13	3	8	10
16	2	5	11

1.   *Dorothy*   *Demeter*   *Miscarriage Not Magic*

Miscarriage Not Magic

15	12	6	1
2	5	11	16
3	8	10	13
14	9	7	4

3.   *Deborah*   *Elara*

12	6	1	15
9	7	4	14
8	10	13	3
5	11	16	2

2.   *Daisy*   *Demeter*   *Miscarriage Not Magic*

Miscarriage Not Magic

6	1	15	12
11	16	2	5
10	13	3	8
7	4	14	9

4.   *Diane*   *Elara*

Twins

Candy of the Hestia family has four daughters, two of the Demeter family and two of the Elara family.

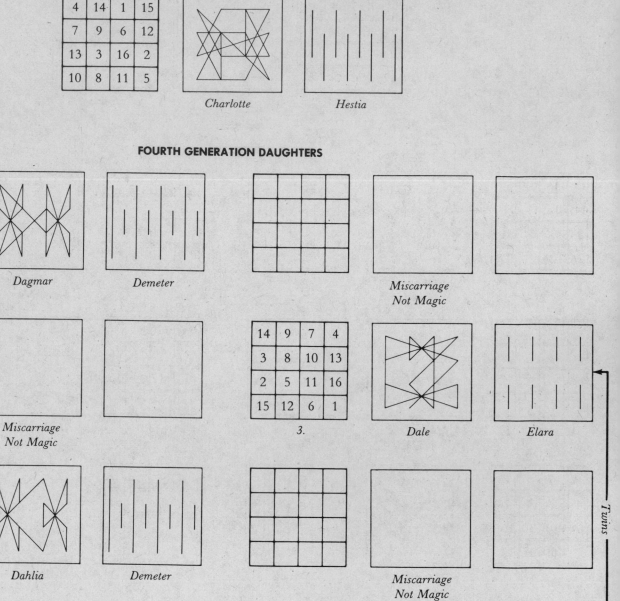

4	14	1	15
7	9	6	12
13	3	16	2
10	8	11	5

Charlotte

Hestia

FOURTH GENERATION DAUGHTERS

4	14	9	7
1	15	12	6
16	2	5	11
13	3	8	10

1.

Dagmar

Demeter

*Miscarriage
Not Magic*

*Miscarriage
Not Magic*

14	9	7	4
3	8	10	13
2	5	11	16
15	12	6	1

3.

Dale

Elara

9	7	4	14
12	6	1	15
5	11	16	2
8	10	13	3

2.

Dahlia

Demeter

*Miscarriage
Not Magic*

Twins

*Miscarriage
Not Magic*

7	4	14	9
10	13	3	8
11	16	2	5
6	1	15	12

4.

Danielle

Elara

Charlotte of the Hestia family has four daughters, two of the Demeter family
and two of the Elara family.

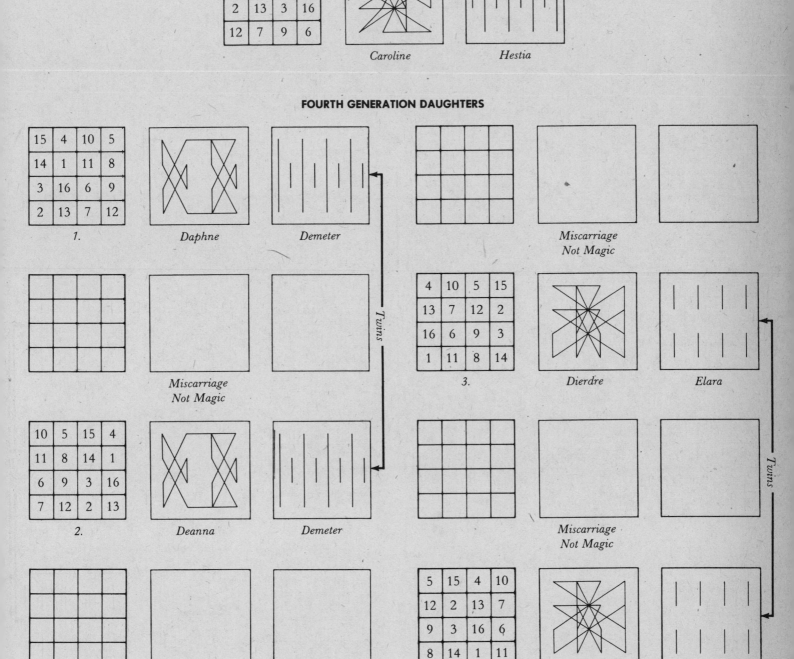

Caroline

Hestia

FOURTH GENERATION DAUGHTERS

1.

Daphne

Demeter

Miscarriage Not Magic

Miscarriage Not Magic

Twins

3.

Dierdre

Elara

2.

Deanna

Demeter

Miscarriage Not Magic

Twins

Miscarriage Not Magic

4.

Delia

Elara

Miscarriage Not Magic

Caroline of the Hestia family has four daughters, two of the Demeter family and two of the Elara family. Note two sets of twins.

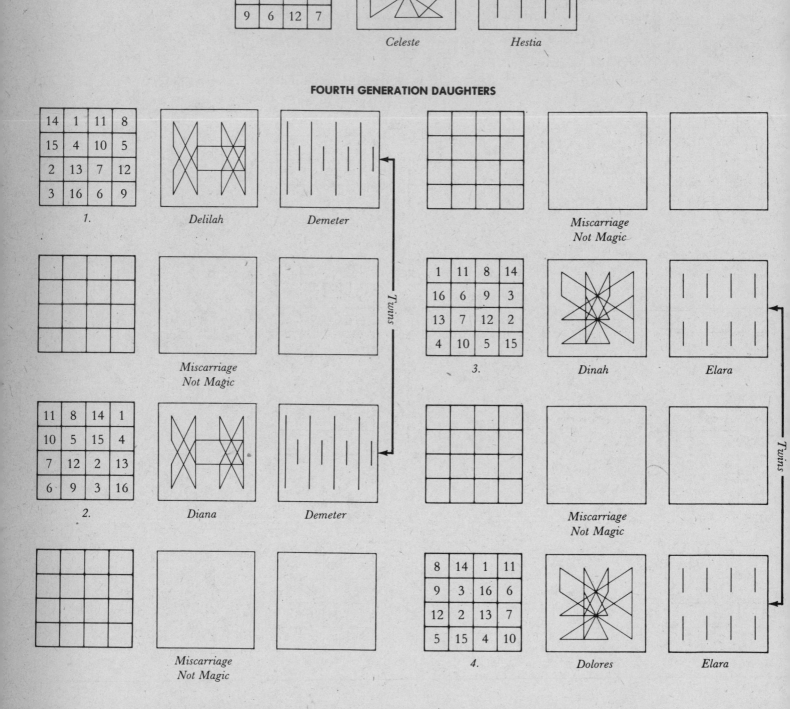

14	1	15	4
8	11	5	10
3	16	2	13
9	6	12	7

Celeste   *Hestia*

FOURTH GENERATION DAUGHTERS

14	1	11	8
15	4	10	5
2	13	7	12
3	16	6	9

1.   *Delilah*   *Demeter*   *Miscarriage Not Magic*

Miscarriage Not Magic   *Twins*

1	11	8	14
16	6	9	3
13	7	12	2
4	10	5	15

3.   *Dinah*   *Elara*

11	8	14	1
10	5	15	4
7	12	2	13
6	9	3	16

2.   *Diana*   *Demeter*   *Miscarriage Not Magic*   *Twins*

Miscarriage Not Magic

8	14	1	11
9	3	16	6
12	2	13	7
5	15	4	10

4.   *Dolores*   *Elara*

Celeste of the Hestia family has four daughters, two of the Demeter family and two of the Elara family. Note two sets of twins.

Some observations regarding the fifth generation (see Charts C1–C4):

1. Charts C1–C4 cover only one branch (or one thirty-second) of the fifth generation.
2. The family, Asteria (last to appear of the six fertile families), appears for the first time on Charts C1 and C2.
3. There is one set of R.R. twins in each group of four daughters.
4. Daughter 1 on Chart C1 is *identical to Queen Mother Anne*.
5. Daughters 1,2,3,4 on Chart C1 are identical cousins of daughters 1,2,3,4 on Chart C2.
6. Daughters 1,2,3,4 on Chart C3 are R.R. cousins of daughters 1,2,3,4 on Chart C4.

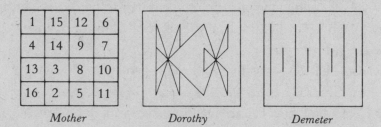

1	15	12	6
4	14	9	7
13	3	8	10
16	2	5	11

Mother   *Dorothy*   *Demeter*

FIFTH GENERATION DAUGHTERS

1	15	14	4
12	6	7	9
8	10	11	5
13	3	2	16

1.   *Electra*   *Niobe*

15	14	4	1
3	2	16	13
10	11	5	8
6	7	9	12

3.   *Emma*   *Asteria*

14	4	1	15
7	9	12	6
11	5	8	10
2	16	13	3

2.   *Ellen*   *Niobe*

4	1	15	14
16	13	3	2
5	8	10	11
9	12	6	7

4.   *Esther*   *Asteria*

Twins

Dorothy of the Demeter family has four daughters, two of the Niobe family
and two of the Asteria family. Note one set of twins.

CHART C2 FOURTH GENERATION Daisy, Second Daughter of Candy

12	6	1	15
9	7	4	14
8	10	13	3
5	11	16	2

Mother

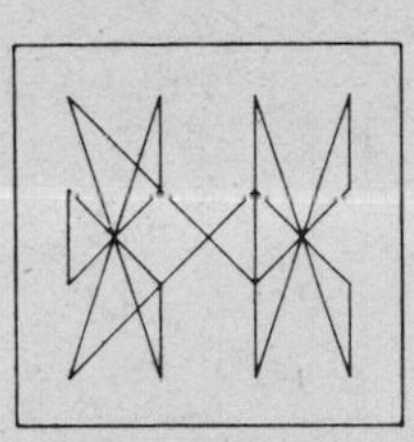

Daisy

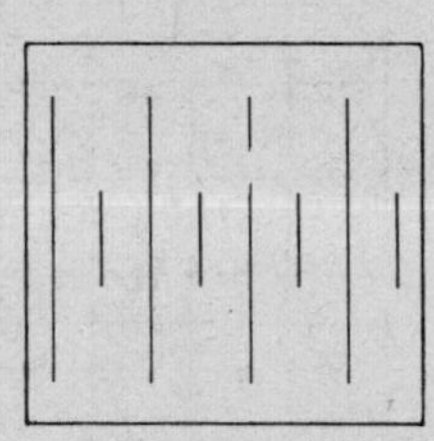

Demeter

12	6	7	9
1	15	14	4
13	3	2	16
8	10	11	5

1.

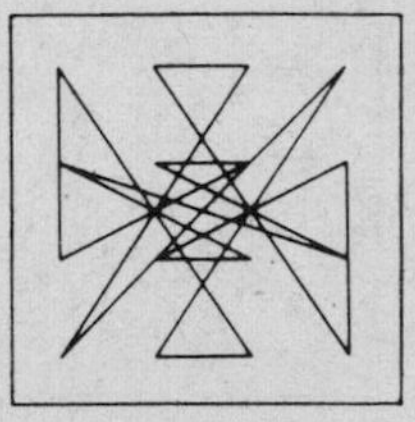

Ella

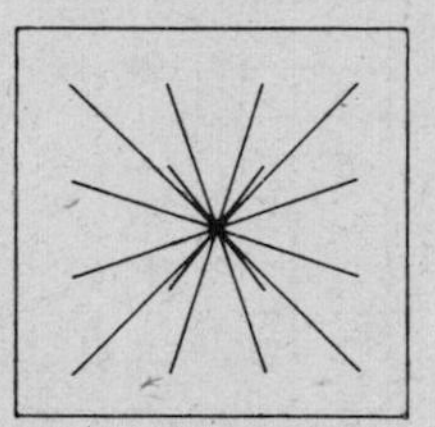

Niobe

6	7	9	12
10	11	5	8
3	2	16	13
15	14	4	1

3.

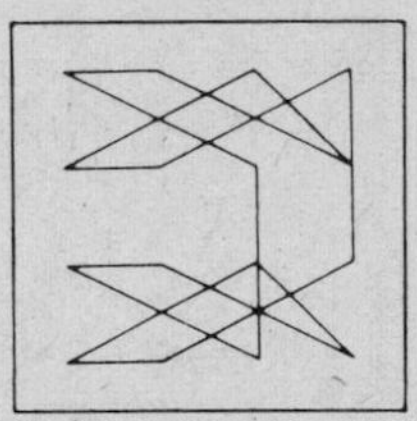

Erica

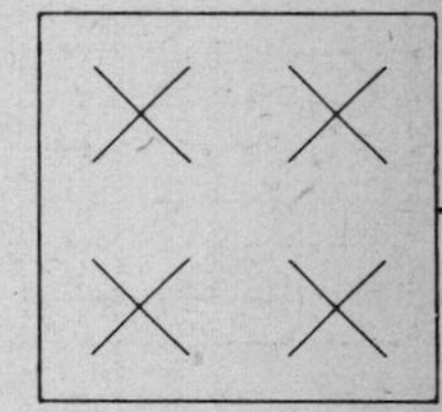

Asteria

7	9	12	6
14	4	1	15
2	16	13	3
11	5	8	10

2.

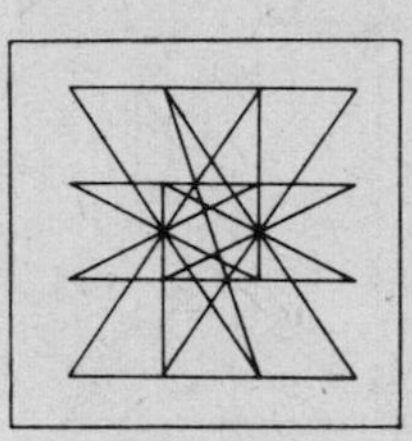

Elizabeth

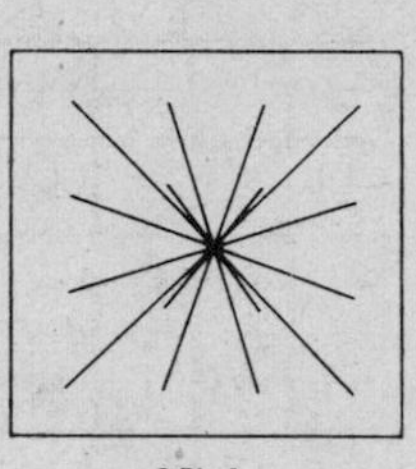

Niobe

9	12	6	7
5	8	10	11
16	13	3	2
4	1	15	14

4.

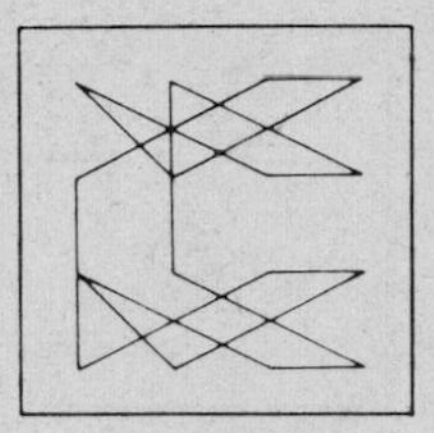

Ethel

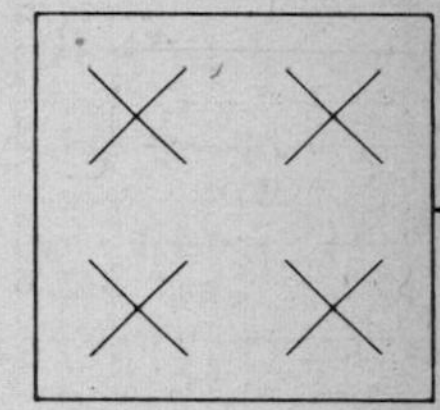

Asteria

Twins

Daisy of the Demeter family has four daughters, two of the Niobe family and two of the Asteria family. Note one set of twins.

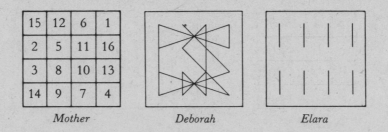

15	12	6	1
2	5	11	16
3	8	10	13
14	9	7	4

Mother   *Deborah*   *Elara*

FIFTH GENERATION DAUGHTERS

15	12	5	2
6	1	16	11
10	13	4	7
3	8	9	14

1.   *Essie*   *Demeter*

12	5	2	15
8	9	14	3
13	4	7	10
1	16	11	6

3.   *Eko*   *Elara*

5	2	15	12
16	11	6	1
4	7	10	13
9	14	3	8

2.   *Elinor*   *Demeter*

2	15	12	5
14	3	8	9
7	10	13	4
11	6	1	16

4.   *Eto*   *Elara*

Twins

Deborah of the Elara family has four daughters, two of the Demeter family
and two of the Elara family. Note one set of twins.

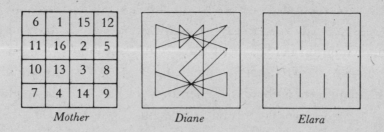

CHART C4 **FOURTH GENERATION** **Diane, Fourth Daughter of Candy**

6	1	15	12
11	16	2	5
10	13	3	8
7	4	14	9

Mother   *Diane*   *Elara*

FIFTH GENERATION DAUGHTERS

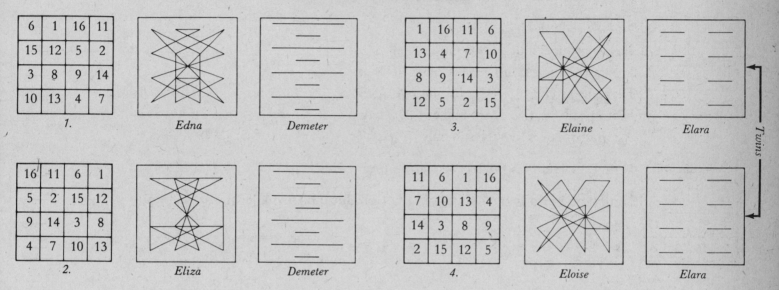

6	1	16	11
15	12	5	2
3	8	9	14
10	13	4	7

1.   *Edna*   *Demeter*

1	16	11	6
13	4	7	10
8	9	14	3
12	5	2	15

3.   *Elaine*   *Elara*

16	11	6	1
5	2	15	12
9	14	3	8
4	7	10	13

2.   *Eliza*   *Demeter*

11	6	1	16
7	10	13	4
14	3	8	9
2	15	12	5

4.   *Eloise*   *Elara*

Twins

Diane of the Elara family has four daughters, two of the Demeter family and two of the Elara family. Note one set of twins.

Some observations regarding the sixth generation (see Charts D2, D3, and D4):

1. Chart D1 has been omitted because it would be a mere duplicate of Chart A.
2. Charts D2, D3, and D4 cover only a small fraction of the sixth generation.
3. Like Queen Mother Anne, who was also of the Niobe family, Mother Ellen bears eight daughters of the Hera family, including two sets of twins (see Chart D2).
4. Ellen's eight daughters of the sixth generation are R.R. images of Queen Mother Anne's eight daughters of the second generation. (Compare Charts D2 and A.)
5. Charts D3 and D4 each show two sets of twins, all of Hestia family.

Some observations regarding families:

1. Niobe produces eight Hera only.
2. Hera produces four Hestia only.
3. Hestia produces two Demeter and two Elara.
4. Demeter produces two Niobe and two Asteria.
5. Elara produces two Elara and two Demeter.
6. Asteria produces four Hestia only.

Note that Hera and Asteria each produce 4 Hestia.

Note that Elara is the only family that produces offspring of the same family.

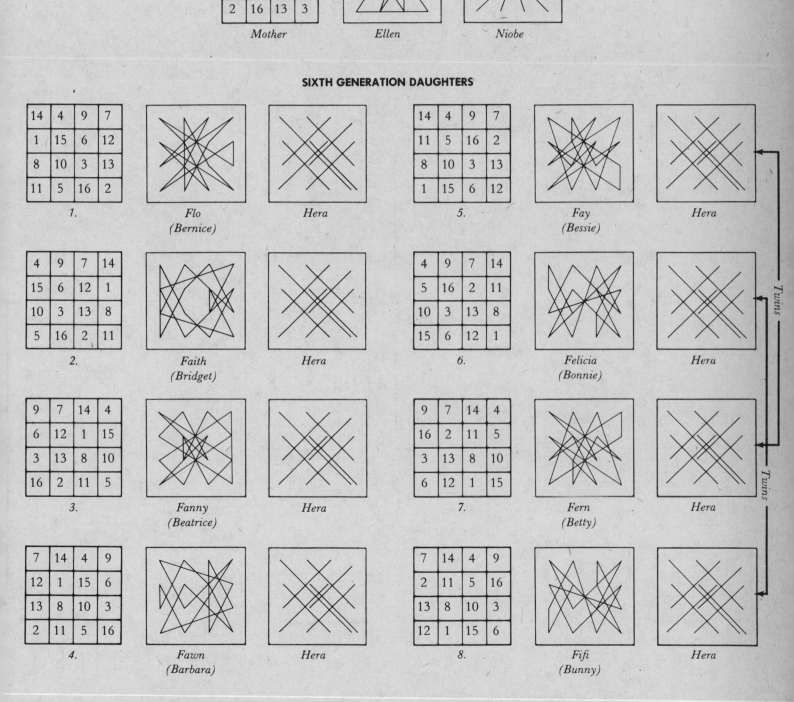

14	4	1	15
7	9	12	6
11	5	8	10
2	16	13	3

Mother   *Ellen*   *Niobe*

SIXTH GENERATION DAUGHTERS

14	4	9	7
1	15	6	12
8	10	3	13
11	5	16	2

1.   *Flo*   *Hera*
                (Bernice)

14	4	9	7
11	5	16	2
8	10	3	13
1	15	6	12

5.   *Fay*   *Hera*
                (Bessie)

4	9	7	14
15	6	12	1
10	3	13	8
5	16	2	11

2.   *Faith*   *Hera*
                (Bridget)

4	9	7	14
5	16	2	11
10	3	13	8
15	6	12	1

6.   *Felicia*   *Hera*
                (Bonnie)

9	7	14	4
6	12	1	15
3	13	8	10
16	2	11	5

3.   *Fanny*   *Hera*
                (Beatrice)

9	7	14	4
16	2	11	5
3	13	8	10
6	12	1	15

7.   *Fern*   *Hera*
                (Betty)

7	14	4	9
12	1	15	6
13	8	10	3
2	11	5	16

4.   *Fawn*   *Hera*
                (Barbara)

7	14	4	9
2	11	5	16
13	8	10	3
12	1	15	6

8.   *Fifi*   *Hera*
                (Bunny)

Twins

Twins

Ellen of the Niobe family has eight daughters, all of the Hera family. Note two sets of twins.

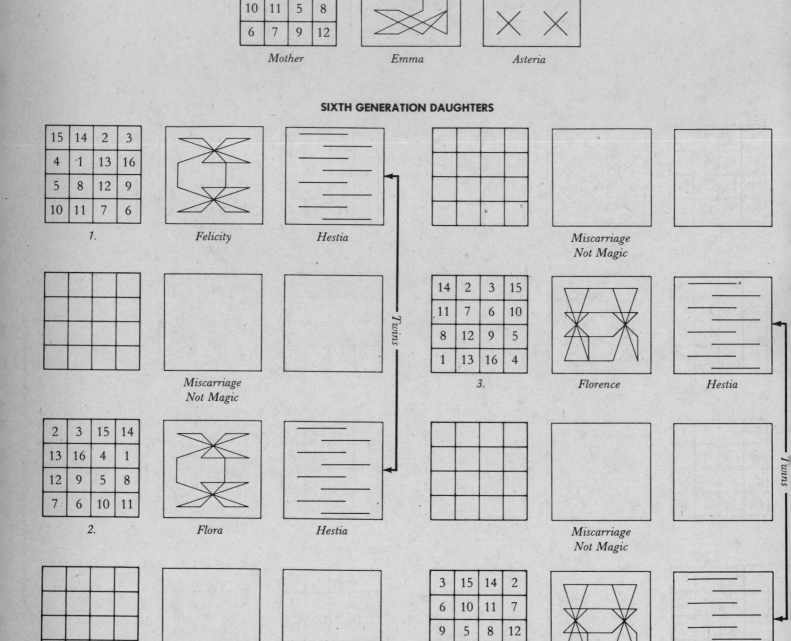

Mother

15	14	4	1
3	2	16	13
10	11	5	8
6	7	9	12

Emma

Asteria

SIXTH GENERATION DAUGHTERS

15	14	2	3
4	1	13	16
5	8	12	9
10	11	7	6

1.   *Felicity*   *Hestia*

Miscarriage
Not Magic

Miscarriage
Not Magic

14	2	3	15
11	7	6	10
8	12	9	5
1	13	16	4

3.   *Florence*   *Hestia*

2	3	15	14
13	16	4	1
12	9	5	8
7	6	10	11

2.   *Flora*   *Hestia*

Twins

Twins

Miscarriage
Not Magic

Miscarriage
Not Magic

3	15	14	2
6	10	11	7
9	5	8	12
16	4	1	13

4.   *Francis*   *Hestia*

Emma of the Asteria family has four daughters, all of the Hestia family. Note two sets of twins.

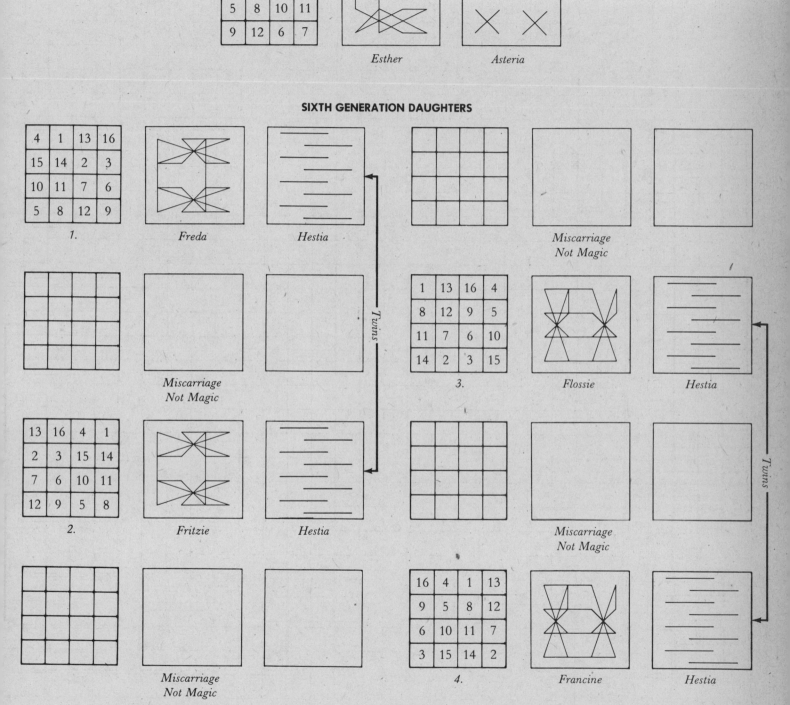

4	1	15	14
16	13	3	2
5	8	10	11
9	12	6	7

Esther   *Asteria*

SIXTH GENERATION DAUGHTERS

4	1	13	16
15	14	2	3
10	11	7	6
5	8	12	9

1.   *Freda*   *Hestia*

Miscarriage Not Magic

Twins

1	13	16	4
8	12	9	5
11	7	6	10
14	2	3	15

3.   *Flossie*   *Hestia*

Miscarriage Not Magic

13	16	4	1
2	3	15	14
7	6	10	11
12	9	5	8

2.   *Fritzie*   *Hestia*

Miscarriage Not Magic

Twins

16	4	1	13
9	5	8	12
6	10	11	7
3	15	14	2

4.   *Francine*   *Hestia*

Miscarriage Not Magic

Esther of the Asteria family has four daughters, all of the Hestia family. Note two sets of twins.

METHOD A FOR 8 X 8

Method A requires a preliminary square.

Instructions for constructing preliminary square:

1. Fill in 8 × 8 diagram with numbers 1 through 64 in their natural order and draw
dashed diagonal lines as indicated.

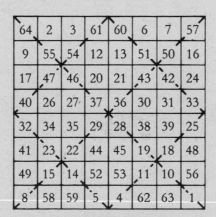

Preliminary Square

Instructions for constructing final magic square:

1. Each number cut by the diagonal lines of preliminary square is replaced by its
complementary number in the final magic square.
2. Numbers not cut by diagonal lines of preliminary square are placed in the final
magic square in their original positions.

64	2	3	61	60	6	7	57
9	55	54	12	13	51	50	16
17	47	46	20	21	43	42	24
40	26	27	37	36	30	31	33
32	34	35	29	28	38	39	25
41	23	22	44	45	19	18	48
49	15	14	52	53	11	10	56
8	58	59	5	4	62	63	1

Final Magic Square

METHOD B FOR 8 X 8 :
EIGHT VARIATIONS OF INTERCHANGE

AS EXPLAINED BY MAURICE KRAITCHIK

64	2	3	61	60	6	7	57
9	55	54	12	13	51	50	16
17	47	46	20	21	43	42	24
40	26	27	37	36	30	31	33
32	34	35	29	28	38	39	25
41	23	22	44	45	19	18	48
49	15	14	52	53	11	10	56
8	58	59	5	4	62	63	1

Original Magic Square

8	58	59	5	4	62	63	1
9	55	54	12	13	51	50	16
17	47	46	20	21	43	42	24
40	26	27	37	36	30	31	33
32	34	35	29	28	38	39	25
41	23	22	44	45	19	18	48
49	15	14	52	53	11	10	56
64	2	3	61	60	6	7	57

Magic Square 1
Original with Border Rows Interchanged

57	2	3	61	60	6	7	64
16	55	54	12	13	51	50	9
24	47	46	20	21	43	42	17
33	26	27	37	36	30	31	40
25	34	35	29	28	38	39	32
48	23	22	44	45	19	18	41
56	15	14	52	53	11	10	49
1	58	59	5	4	62	63	8

Magic Square 2
Original with Border Columns Interchanged

1	58	59	5	4	62	63	8
16	55	54	12	13	51	50	9
24	47	46	20	21	43	42	17
33	26	27	37	36	30	31	40
25	34	35	29	28	38	39	32
48	23	22	44	45	19	18	41
56	15	14	52	53	11	10	49
57	2	3	61	60	6	7	64

Magic Square 3
Original with Border Rows Interchanged,
Then Border Columns Interchanged

28	38	39	25	32	34	35	29
45	19	18	48	41	23	22	44
53	11	10	56	49	15	14	52
4	62	63	1	8	58	59	5
60	6	7	57	64	2	3	61
13	51	50	16	9	55	54	12
21	43	42	24	17	47	46	20
36	30	31	33	40	26	27	37

Magic Square 4
Original with Quadrants Interchanged

64	2	3	61	60	6	7	57
9	55	54	12	13	51	50	16
17	47	46	20	21	43	42	24
40	26	27	37	36	30	31	33
32	34	35	29	28	38	39	25
41	23	22	44	45	19	18	48
49	15	14	52	53	11	10	56
8	58	59	5	4	62	63	1

Original Magic Square

8	58	59	5	4	62	63	1
49	15	14	52	53	11	10	56
17	47	46	20	21	43	42	24
40	26	27	37	36	30	31	33
32	34	35	29	28	38	39	25
41	23	22	44	45	19	18	48
9	55	54	12	13	51	50	16
64	2	3	61	60	6	7	57

Magic Square 5
Original with Two Top Rows Interchanged
with Two Bottom Rows

57	7	3	61	60	6	2	64
16	50	54	12	13	51	55	9
24	42	46	20	21	43	47	17
33	31	27	37	36	30	26	40
25	39	35	29	28	38	34	32
48	18	22	44	45	19	23	41
56	10	14	52	53	11	15	49
1	63	59	5	4	62	58	8

Magic Square 6
Original with Two Left Columns Interchanged
with Two Right Columns

1	63	59	5	4	62	58	8
56	10	14	52	53	11	15	49
24	42	46	20	21	43	47	17
33	31	27	37	36	30	26	40
25	39	35	29	28	38	34	32
48	18	22	44	45	19	23	41
16	50	54	12	13	51	55	9
57	7	3	61	60	6	2	64

Magic Square 7
Original with Two Top Rows Interchanged
with Two Bottom Rows, Then Two Left Columns
Interchanged with Two Right Columns

28	38	34	32	25	39	35	29
45	19	23	41	48	18	22	44
13	51	55	9	16	50	54	12
60	6	2	64	57	7	3	61
4	62	58	8	1	63	59	5
53	11	15	49	56	10	14	52
21	43	47	17	24	42	46	20
36	30	26	40	33	31	27	37

Magic Square 8
Magic Square 7 (See Left) with Quadrants Interchanged

Project: Can you devise other interchange systems? There are others for 8 × 8.

METHOD C FOR 8 × 8

INVENTED BY FREDERICK A. WOODRUFF

This method requires one preliminary square and two 2 × 4 magic rectangles.

Instructions for constructing preliminary square and first magic rectangle:

1. Construct 8 × 8 diagram. Draw a vertical line down the middle and add an extra row across top and bottom as indicated by dashed lines in figure 1.
2. Construct a 2 × 4 magic rectangle by placing the numbers 1 through 8 in a 2 × 4 diagram in any order, provided vertical pairs add up to 9 and both rows add up to 18, as shown in figure 2.
3. Write the row numbers of magic rectangle (figure 2) alternately at the top and bottom of the eight columns of figure 1, as shown by dashed lines.

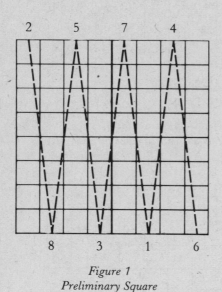

2	8	5	3
7	1	4	6

Figure 1
Preliminary Square

Figure 2

	2		5		7		4	
1	9	64	33	24	49	8	25	48	(8)
2	10	63	34	23	50	7	26	47	(7)
3	11	62	35	22	51	6	27	46	(6)
4	12	61	36	21	52	5	28	45	(5)
5	13	60	37	20	53	4	29	44	(4)
6	14	59	38	19	54	3	30	43	(3)
7	15	58	39	18	55	2	31	42	(2)
8	16	57	40	17	56	1	32	41	(1)
	8		3		1		6		

Figure 3
Preliminary Square

4. Each column now bears a number. Following the arithmetical order of the numbered columns, write in the numbers 1 through 64 in groups of eight upward and downward as shown in figure 3.
5. Starting from top, write index numbers 1 through 8 on the outside of preliminary square to the left of column one, as shown in figure 3. Starting from bottom, write the numbers one through eight outside and to the right of column eight as shown in figure 3. *Circle these numbers.*

This completes the construction of the preliminary square.

Instructions for construction of final 8 × 8 magic square:

1. Construct another 2 × 4 magic rectangle with numbers in different arrangement from first (see figure 4).

1	7	6	4
8	2	3	5

Figure 4

2. Construct an 8 × 8 diagram and draw a vertical line down middle.
3. Use the four numbers of the top row of the magic rectangle (figure 4) and place these numbers running downward to the left of column 1, as shown in figure 5.

1	9	64	33	24	17	40	57	16	8
(5)	52	5	28	45	44	29	4	53	(4)
7	15	58	39	18	23	34	63	10	2
(3)	54	3	30	43	46	27	6	51	(6)
6	14	59	38	19	22	35	62	11	3
(2)	55	2	31	42	47	26	7	50	(7)
4	12	61	36	21	20	37	60	13	5
(8)	49	8	25	48	41	32	1	56	(1)

Figure 5

4. Place and *circle* the same numbers running upward in alternate boxes to right of column eight, as shown in figure 5.
5. Numbers of bottom row of rectangle are placed in similar manner (see figure 5).
6. Fill in the 64 boxes of figure 5 with numbers in groups of four as follows:
   (a) Observe that the uncircled number 1 has been placed to the left of the first four boxes in top row of figure 5. Using this number as an index, refer to figure 3 and note that the four numbers to the right of the uncircled number 1 (to left of column one) are 9, 64, 33, and 24. Place these four numbers in the four boxes to the right of the uncircled 1 in figure 5.
   (b) Observe that circled number 5 is to left of first four boxes of second row of figure 5. Locate the circled 5 in figure 3 and note that it indexes 52, 5, 38, 45. Place these numbers in figure 5.
   (c) Observe uncircled 7 in figure 5. Locate uncircled 7 in figure 3 and note that it indexes number 15, 58, 39, 18. Place these numbers to right of uncircled 7 in figure 5.

(d) Proceeding downward, fill in the first four boxes of the remaining rows of figure 5 in same manner.

(e) The second half of figure 5, beginning with second four boxes of row one, is completed as before, except that the order of each group of four indexed numbers is *reversed*.

Note: Each different arrangement of the numbers in either of the magic rectangles will produce a different magic square.

Magic squares constructed by this method are associative and pandiagonal.

The 8 × 8 magic square shown below was constructed by 8 × 8 method C. Any four numbers enclosed by a 2 × 2 boundary line will total 130 (one-half of constant 260). A convenient way to show this is by use of a cutout card. Cut 2 × 2 hole in card and place over the 8 × 8 magic square. Any four numbers showing through the window will total 130 (see below).

This feature is true of all magic squares made by 8 × 8 method C.

2	31	42	55	50	47	26	7
64	33	24	9	16	17	40	57
3	30	43	54	51	46	27	6
61	36	21	12	13	20	37	60
5	28	45	52	53	44	29	4
59	38	19	14	11	22	35	62
8	25	48	49	56	41	32	1
58	39	18	15	10	23	34	63

Total 130
Total 130
Total 130
Total 130
Total 130

TWELVE 8 X 8 SEQUENCE DESIGNS

The following twelve different sequence designs were based on the 8 × 8 magic squares shown below each design. They were made by drawing a straight line from numbers 1 to 2 to 3 to 4 to 5 and so forth, with a final line drawn from 64 to 1.

These and a great variety of other designs based on magic squares offer rich possibilities to painters, sculptors, architects, and decorators, as well as to designers of fabric, rugs, tile, tapestry, wallpaper, and string structures.

All designs made by this method present interesting repeat patterns when arranged in tile formation, as shown on pages 157–168.

Note: The preliminary squares shown below in designs 1 through 11 were constructed by use of the natural sequence of numbers.

The preliminary square shown below in design 12 was constructed by use of a complementary sequence.

	2		5		7		4		
1	9	64	33	24	49	8	25	48	8
2	10	63	34	23	50	7	26	47	7
3	11	62	35	22	51	6	27	46	6
4	12	61	36	21	52	5	28	45	5
5	13	60	37	20	53	4	29	44	4
6	14	59	38	19	54	3	30	43	3
7	15	58	39	18	55	2	31	42	2
8	16	57	40	17	56	1	32	41	1
		8		3		1		6	

2	8	5	3
7	1	4	6

Preliminary Square Using Natural Sequence

1	9	64	33	24	17	40	57	16	8
5	52	5	28	45	44	29	4	53	4
7	15	58	39	18	23	34	63	10	2
3	54	3	30	43	46	27	6	51	6
6	14	59	38	19	22	35	62	11	3
2	55	2	31	42	47	26	7	50	7
4	12	61	36	21	20	37	60	13	5
8	49	8	25	48	41	32	1	56	1

1	7	6	4
8	2	3	5

Magic Square

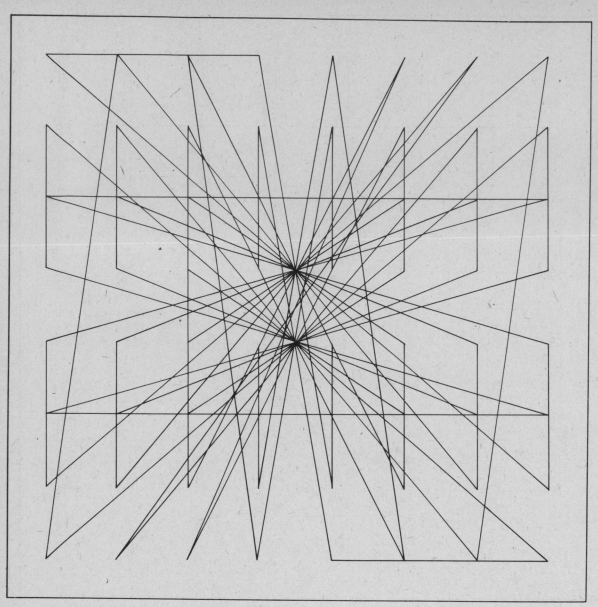

Figure 2

	6		1		3		8		
1	41	32	1	56	17	40	57	16	⑧
2	42	31	2	55	18	39	58	15	⑦
3	43	30	3	54	19	38	59	14	⑥
4	44	29	4	53	20	37	60	13	⑤
5	45	28	5	52	21	36	61	12	④
6	46	27	6	51	22	35	62	11	③
7	47	26	7	50	23	34	63	10	②
8	48	25	8	49	24	33	64	9	①
		4		7		5		2	

Preliminary Square Using Natural Sequence

6	4	1	7
3	5	8	2

8	48	25	8	49	56	1	32	41	1
⑦	18	39	58	15	10	63	34	23	②
5	45	28	5	52	53	4	29	44	4
⑥	19	38	59	14	11	62	35	22	③
3	43	30	3	54	51	6	27	46	6
④	21	36	61	12	13	60	37	20	⑤
2	42	31	2	55	50	7	26	47	7
①	24	33	64	9	16	57	40	17	⑧

Magic Square

8	5	3	2
1	4	6	7

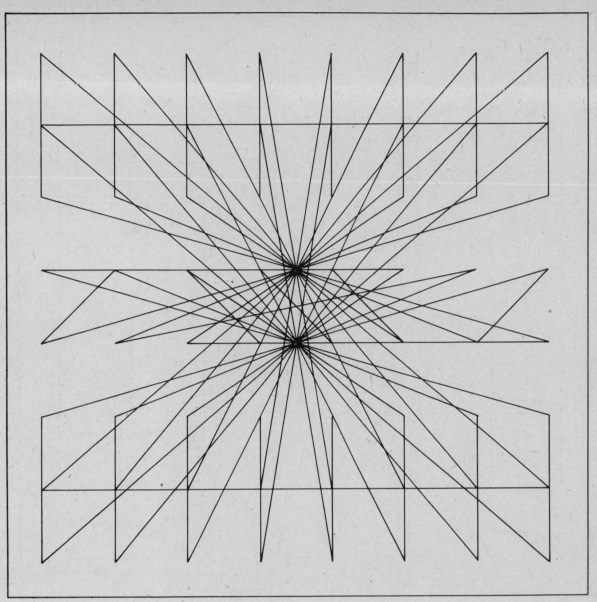

Figure 3

	7		6		2		3		
1	49	8	41	32	9	64	17	40	(8)
2	50	7	42	31	10	63	18	39	(7)
3	51	6	43	30	11	62	19	38	(6)
4	52	5	44	29	12	61	20	37	(5)
5	53	4	45	28	13	60	21	36	(4)
6	54	3	46	27	14	59	22	35	(3)
7	55	2	47	26	15	58	23	34	(2)
8	56	1	48	25	16	57	24	33	(1)
		1		4		8		5	

7	1	6	4
2	8	3	5

Preliminary Square Using Natural Sequence

3	2	8	5
6	7	1	4

3	51	6	43	30	27	46	3	54	6
(4)	13	60	21	36	37	20	61	12	(5)
2	50	7	42	31	26	47	2	55	7
(1)	16	57	24	33	40	17	64	9	(8)
8	56	1	48	25	32	41	8	49	1
(7)	10	63	18	39	34	23	58	15	(2)
5	53	4	45	28	29	44	5	52	4
(6)	11	62	19	38	35	22	59	14	(3)

Magic Square

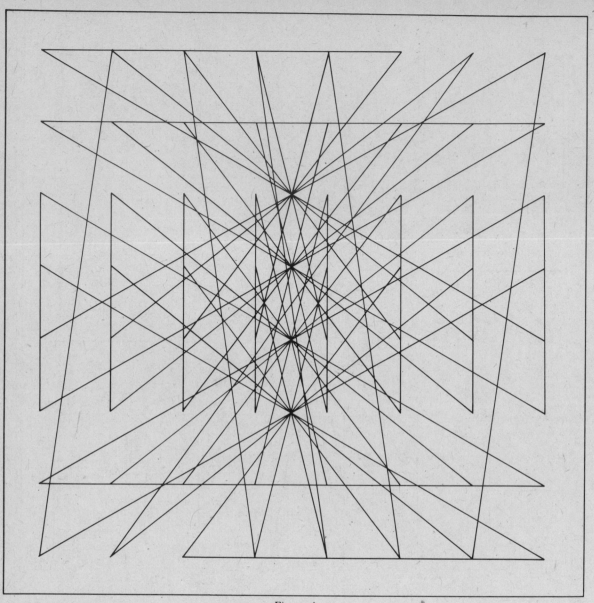

Figure 4

Preliminary Square Using Natural Sequence:

	6		7		3		2		
1	41	32	49	8	17	40	9	64	(8)
2	42	31	50	7	18	39	10	63	(7)
3	43	30	51	6	19	38	11	62	(6)
4	44	29	52	5	20	37	12	61	(5)
5	45	28	53	4	21	36	13	60	(4)
6	46	27	54	3	22	35	14	59	(3)
7	47	26	55	2	23	34	15	58	(2)
8	48	25	56	1	24	33	16	57	(1)
		4		1		5		8	

Preliminary Square Using Natural Sequence

6	4	7	1
3	5	2	8

Magic Square:

	8								
8	48	25	56	1	8	49	32	41	1
(4)	21	36	13	60	61	12	37	20	(5)
3	43	30	51	6	3	54	27	46	6
(7)	18	39	10	63	58	15	34	23	(2)
2	42	31	50	7	2	55	26	47	7
(6)	19	38	11	62	59	14	35	22	(3)
5	45	28	53	4	5	52	29	44	4
(1)	24	33	16	57	64	9	40	17	(8)

Magic Square

8	3	2	5
1	6	7	4

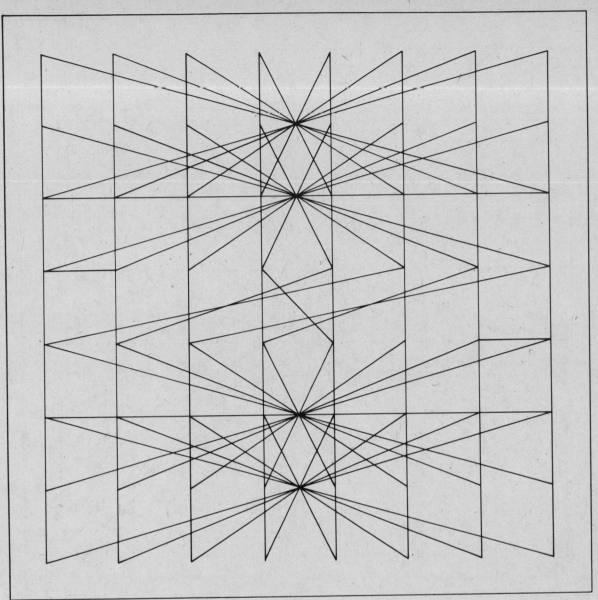

Figure 5

Preliminary Square Using Natural Sequence

	7		1		2		8		
1	49	48	1	32	9	24	64	40	⑧
2	50	47	2	31	10	23	63	39	⑦
3	51	46	3	30	11	22	62	38	⑥
4	52	45	4	29	12	21	61	37	⑤
5	53	44	5	28	13	20	60	36	④
6	54	43	6	27	14	19	59	35	③
7	55	42	7	26	15	18	58	34	②
8	56	41	8	25	16	17	57	33	①
		6		4		3		5	

7	6	1	4
2	3	8	5

3	5	8	2
6	4	1	7

Magic Square

3	51	46	3	30	27	6	43	54	6
⑦	10	23	58	39	34	63	18	15	②
5	53	44	5	28	29	4	45	52	4
①	16	17	64	33	40	57	24	9	⑧
8	56	41	8	25	32	1	48	49	1
④	13	20	61	36	37	60	21	12	⑤
2	50	47	2	31	26	7	42	55	7
⑥	11	22	59	38	35	62	19	14	③

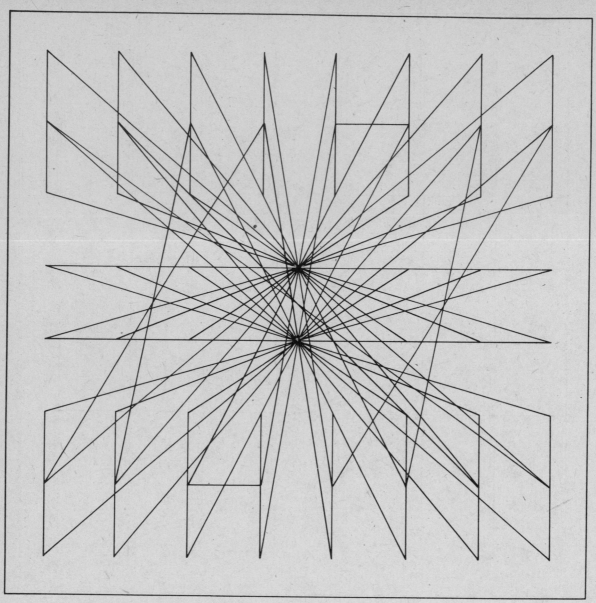

Figure 6

	1		6		8		3		
1	1	32	41	56	57	40	17	16	⑧
2	2	31	42	55	58	39	18	15	⑦
3	3	30	43	54	59	38	19	14	⑥
4	4	29	44	53	60	37	20	13	⑤
5	5	28	45	52	61	36	21	12	④
6	6	27	46	51	62	35	22	11	③
7	7	26	47	50	63	34	23	10	②
8	8	25	48	49	64	33	24	9	①
		4		7		5		2	

Preliminary Square Using Natural Sequence

1	4	6	7
8	5	3	2

2	3	5	8
7	6	4	1

2	2	31	42	55	50	47	26	7	7
①	64	33	24	9	16	17	40	57	⑧
3	3	30	43	54	51	46	27	6	6
④	61	36	21	12	13	20	37	60	⑤
5	5	28	45	52	53	44	29	4	4
⑥	59	38	19	14	11	22	35	62	③
8	8	25	48	49	56	41	32	1	1
⑦	58	39	18	15	10	23	34	63	②

Magic Square

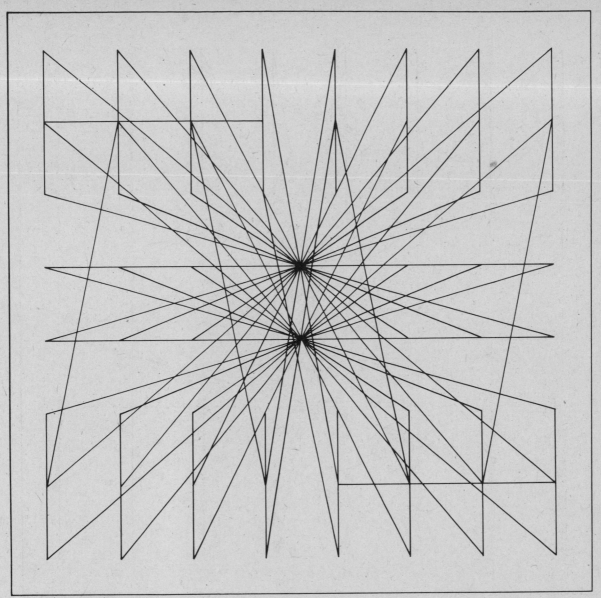

Figure 7

	7		4		2		5		
1	49	8	25	48	9	64	33	24	(8)
2	50	7	26	47	10	63	34	23	(7)
3	51	6	27	46	11	62	35	22	(6)
4	52	5	28	45	12	61	36	21	(5)
5	53	4	29	44	13	60	37	20	(4)
6	54	3	30	43	14	59	38	19	(3)
7	55	2	31	42	15	58	39	18	(2)
8	56	1	32	41	16	57	40	17	(1)
		1		6		8		3	

7	1	4	6
2	8	5	3

Preliminary Square Using Natural Sequence

2	50	7	26	47	42	31	2	55	7
(1)	16	57	40	17	24	33	64	9	(8)
3	51	6	27	46	43	30	3	54	6
(4)	13	60	37	20	21	36	61	12	(5)
5	53	4	29	44	45	28	5	52	4
(6)	11	62	35	22	19	38	59	14	(3)
8	56	1	32	41	48	25	8	49	1
(7)	10	63	34	23	18	39	58	15	(2)

2	3	5	8
7	6	4	1

Magic Square

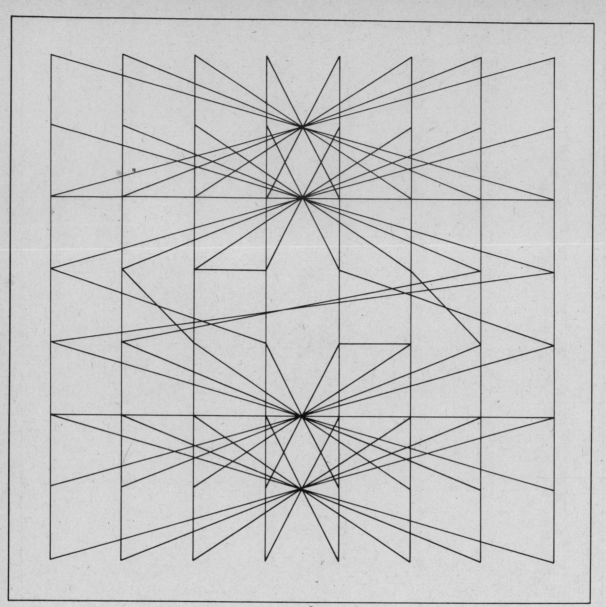

Figure 8

Preliminary Square Using Natural Sequence

	1		6		8		3		
1	1	32	41	56	57	40	17	16	(8)
2	2	31	42	55	58	39	18	15	(7)
3	3	30	43	54	59	38	19	14	(6)
4	4	29	44	53	60	37	20	13	(5)
5	5	28	45	52	61	36	21	12	(4)
6	6	27	46	51	62	35	22	11	(3)
7	7	26	47	50	63	34	23	10	(2)
8	8	25	48	49	64	33	24	9	(1)
		4		7		5		2	

1	4	6	7
8	5	3	2

6	4	1	7
3	5	8	2

Magic Square

	6		4		1		7		
6	6	27	46	51	54	43	30	3	3
(2)	63	34	23	10	15	18	39	58	(7)
4	4	29	44	53	52	45	28	5	5
(8)	57	40	17	16	9	24	33	64	(1)
1	1	32	41	56	49	48	25	8	8
(5)	60	37	20	13	12	21	36	61	(4)
7	7	26	47	50	55	42	31	2	2
(3)	62	35	22	11	14	19	38	59	(6)

Figure 9

	1		6		8		3		
1	1	32	41	56	57	40	17	16	⑧
2	2	31	42	55	58	39	18	15	⑦
3	3	30	43	54	59	38	19	14	⑥
4	4	29	44	53	60	37	20	13	⑤
5	5	28	45	52	61	36	21	12	④
6	6	27	46	51	62	35	22	11	③
7	7	26	47	50	63	34	23	10	②
8	8	25	48	49	64	33	24	9	①
		4		7		5		2	

Preliminary Square Using Natural Sequence

1	4	6	7
8	5	3	2

2	8	5	3
7	1	4	6

2	2	31	42	55	50	47	26	7	7
⑥	59	38	19	14	11	22	35	62	③
8	8	25	48	49	56	41	32	1	1
④	61	36	21	12	13	20	37	60	⑤
5	5	28	45	52	53	44	29	4	4
①	64	33	24	9	16	17	40	57	⑧
3	3	30	43	54	51	46	27	6	6
⑦	58	39	18	15	10	23	34	63	②

Magic Square

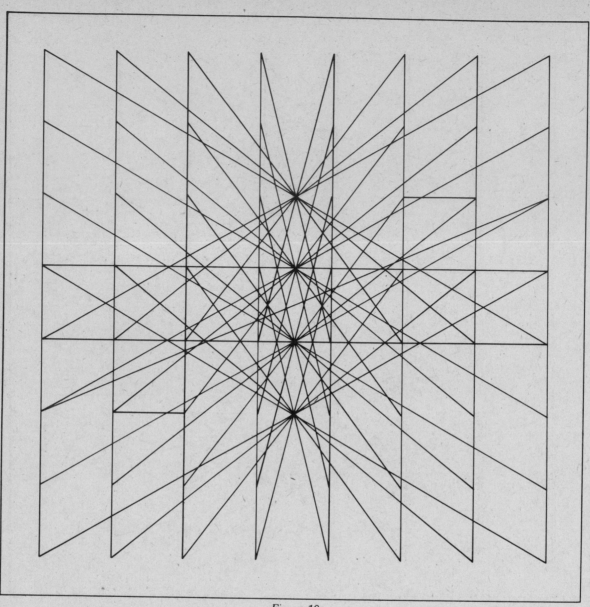

Figure 10

Preliminary Square Using Natural Sequence

	1		7		8		2		
1	1	48	49	32	57	24	9	40	(8)
5	2	47	50	31	58	23	10	39	(4)
3	3	46	51	30	59	22	11	38	(6)
7	4	45	52	29	60	21	12	37	(2)
2	5	44	53	28	61	20	13	36	(7)
6	6	43	54	27	62	19	14	35	(3)
4	7	42	55	26	63	18	15	34	(5)
8	8	41	56	25	64	17	16	33	(1)
		6		4		3		5	

1	6	7	4
8	3	2	5

Magic Square

3	3	46	51	30	27	54	43	6	6
(4)	58	23	10	39	34	15	18	63	(5)
8	8	41	56	25	32	49	48	1	1
(7)	61	20	13	36	37	12	21	60	(2)
2	5	44	53	28	29	52	45	4	7
(1)	64	17	16	33	40	9	24	57	(8)
5	2	47	50	31	26	55	42	7	4
(6)	59	22	11	38	35	14	19	62	(3)

3	8	2	5
6	1	7	4

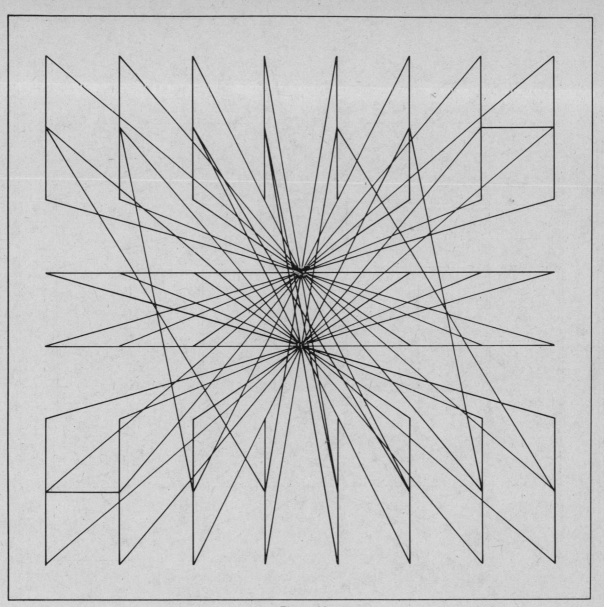

Figure 11

		6		1		3		8		
1	41	56	1	32	17	16	57	40	⑧
2	42	55	2	31	18	15	58	39	⑦
3	43	54	3	30	19	14	59	38	⑥
4	44	53	4	29	20	13	60	37	⑤
5	45	52	5	28	21	12	61	36	④
6	46	51	6	27	22	11	62	35	③
7	47	50	7	26	23	10	63	34	②
8	48	49	8	25	24	9	64	33	①
		7		4		2		5	

Preliminary Square Using Natural Sequence

6	7	1	4
3	2	8	5

2		42	55	2	31	26	7	50	47	7
①		24	9	64	33	40	57	16	17	⑧
3		43	54	3	30	27	6	51	46	6
④		21	12	61	36	37	60	13	20	⑤
5		45	52	5	28	29	4	53	44	4
⑥		19	14	59	38	35	62	11	22	③
8		48	49	8	25	32	1	56	41	1
⑦		18	15	58	39	34	63	10	23	②

Magic Square

2	3	5	8
7	6	4	1

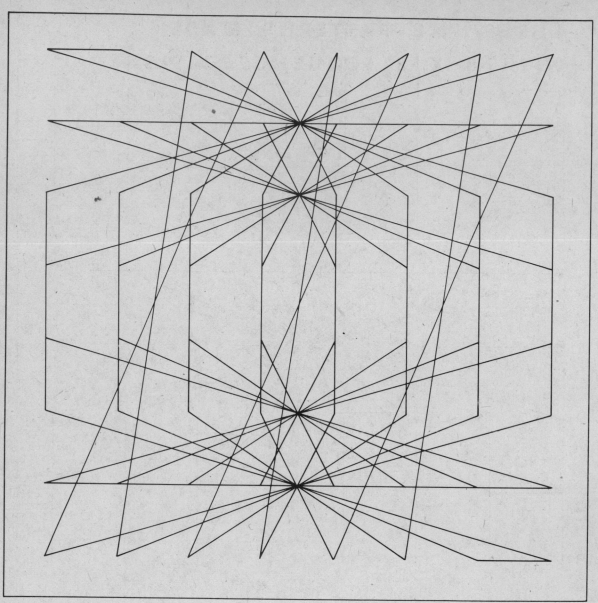

Figure 12

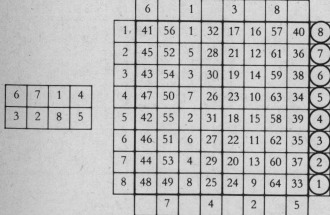

	6		1		3		8		
1	41	56	1	32	17	16	57	40	⑧
2	45	52	5	28	21	12	61	36	⑦
3	43	54	3	30	19	14	59	38	⑥
4	47	50	7	26	23	10	63	34	⑤
5	42	55	2	31	18	15	58	39	④
6	46	51	6	27	22	11	62	35	③
7	44	53	4	29	20	13	60	37	②
8	48	49	8	25	24	9	64	33	①
		7		4		2		5	

Preliminary Square Using Sequence of Complementary Numbers

6	7	1	4
3	2	8	5

8	5	3	2
1	4	6	7

8	48	49	8	25	32	1	56	41	1
⑦	21	12	61	36	37	60	13	20	②
5	42	55	2	31	26	7	50	47	4
⑥	19	14	59	38	35	62	11	22	③
3	43	54	3	30	27	6	51	46	6
④	18	15	58	39	34	63	10	23	⑤
2	45	52	5	28	29	4	53	44	7
①	24	9	64	33	40	57	16	17	⑧

Magic Square

THREE TILE PATTERNS MADE
WITH 8 X 8 SEQUENCE DESIGNS

8 × 8 Sequence Design Reduced in Size and Arranged in a Tile Formation

Six Valuable Substitute Sequences for 8 × 8

Refer to preliminary squares in 8 × 8 methods A and C, and note that they are based on the sequence of numbers 1 through 64 in their natural order. There are other sequence arrangements of these numbers which may be *substituted for the natural sequence*. Each different sequence arrangement will result in a different magic square. Six of these substitute sequences are shown below.

Plus-two sequence:

1	3	2	4	5	7	6	8
9	11	10	12	13	15	14	16
17	19	18	20	21	23	22	24
25	27	26	28	29	31	30	32
33	35	34	36	37	39	38	40
41	43	42	44	45	47	46	48
49	51	50	52	53	55	54	56
57	59	58	60	61	63	62	64

Substitute Plus-Two Sequence for Preliminary Square— 8 × 8 Method A

64	3	2	61	60	7	6	57
9	54	55	12	13	50	51	16
17	46	47	20	21	42	43	24
40	27	26	37	36	31	30	33
32	35	34	29	28	39	38	25
41	22	23	44	45	18	19	48
49	14	15	52	53	10	11	56
8	59	58	5	4	63	62	1

Magic Square Constructed by 8 × 8 Method A Using Plus-Two Sequence in Preliminary Square

Plus-four sequence:

1	5	2	6	3	7	4	8
9	13	10	14	11	15	12	16
17	21	18	22	19	23	20	24
25	29	26	30	27	31	28	32
33	37	34	38	35	39	36	40
41	45	42	46	43	47	44	48
49	53	50	54	51	55	52	56
57	61	58	62	59	63	60	64

Substitute Plus-Four Sequence for Preliminary Square—8 × 8 Method A

64	5	2	59	62	7	4	57
9	52	55	14	11	50	53	16
17	44	47	22	19	42	45	24
40	29	26	35	38	31	28	33
32	37	34	27	30	39	36	25
41	20	23	46	43	18	21	48
49	12	15	54	51	10	13	56
8	61	58	3	6	63	60	1

Magic Square Constructed by 8 × 8 Method A Using Plus-Four Sequence

Plus-eight sequence:

1	9	2	10	3	11	4	12
5	13	6	14	7	15	8	16
17	25	18	26	19	27	20	28
21	29	22	30	23	31	24	32
33	41	34	42	35	43	36	44
37	45	38	46	39	47	40	48
49	57	50	58	51	59	52	60
53	61	54	62	55	63	56	64

Plus-eight sequence used to construct magic square by 8 × 8 method C:

	7		6		2		3		
1	49	12	37	32	5	64	17	44	(8)
2	57	4	45	24	13	56	25	36	(7)
3	50	11	38	31	6	63	18	43	(6)
4	58	3	46	23	14	55	26	35	(5)
5	51	10	39	30	7	62	19	42	(4)
6	59	2	47	22	15	54	27	34	(3)
7	52	9	40	29	8	61	20	41	(2)
8	60	1	48	21	16	53	28	33	(1)
	1		4		8		5		

3	50	11	38	31	22	47	2	59	6
(4)	7	62	19	42	35	26	55	14	(5)
2	57	4	45	24	29	40	9	52	7
(1)	16	53	28	33	44	17	64	5	(8)
8	60	1	48	21	32	37	12	49	1
(7)	13	56	25	36	41	20	61	8	(2)
5	51	10	39	30	23	46	3	58	4
(6)	6	63	18	43	34	27	54	15	(3)

7	1	6	4
2	8	3	5

Preliminary Square

3	2	8	5
6	7	1	4

Final Magic Square

Plus-sixteen sequence:

1	17	2	18	3	19	4	20
5	21	6	22	7	23	8	24
9	25	10	26	11	27	12	28
13	29	14	30	15	31	16	32
33	49	34	50	35	51	36	52
37	53	38	54	39	55	40	56
41	57	42	58	43	59	44	60
45	61	46	62	47	63	48	64

Plus-thirty-two sequence:

1	33	2	34	3	35	4	36
5	37	6	38	7	39	8	40
9	41	10	42	11	43	12	44
13	45	14	46	15	47	16	48
17	49	18	50	19	51	20	52
21	53	22	54	23	55	24	56
25	57	26	58	27	59	28	60
29	61	30	62	31	63	32	64

The complementary-numbers sequence, which is made by:

1. Arranging the numbers 1 through 8 in any order providing the complementary numbers are equidistant from center—1,5,3,7,2,6,4,8 or 5,1,3,2,7,6,8,4, etc.
2. Completing the sequence through to 64 in groups of eight, placed in the same relative order as the first eight numbers (see below).

1.	5	3	7	2	6	4	8
9	13	11	15	10	14	12	16
17	21	19	23	18	22	20	24
25	29	27	31	26	30	28	32
33	37	35	39	34	38	36	40
41	45	43	47	42	46	44	48
49	53	51	55	50	54	52	56
57	61	59	63	58	62	60	64

Substitute Complementary Sequence for Preliminary Square—8 × 8 Method A

64	5	3	58	63	6	4	57
9	52	54	15	10	51	53	16
17	44	46	23	18	43	45	24
40	29	27	34	39	30	28	33
32	37	35	26	31	38	36	25
41	20	22	47	42	19	21	48
49	12	14	55	50	11	13	56
8	61	59	2	7	62	60	1

Magic Square Constructed by 8 × 8 Method A Using Complementary Sequence in Preliminary Square

Note that there are five plus sequences (2,4,8,16,32) which may be used as shown above to make a variety of 8 × 8 magic squares. These five sequences are based on the fact that the number 64 can be divided by these five numbers only (2,4,8,16,32).

The complementary numbers sequence system may be used to create a variety of different magic squares by using different 1 through 8 complementary arrangements.

Note: Plus sequences as well as complementary numbers sequences may be applied to *any* even-order square.

How to make a variety of sequence designs based on the same 8 x 8

The following five 8 × 8 sequence designs make use of the five previously explained plus sequences (plus 2, plus 4, plus 8, plus 16, plus 32) which may be used with 8 × 8 squares.

Each of the five sequence designs was derived from *the same magic square* (see the magic square below the sequence design).

Figure 1: This sequence design makes use of the "plus-two" sequence and was made in two parts:

Part one was made by drawing a straight line from 1 to 3 to 5 to 7 and so forth, and a final line from 63 to 1.

Part two was made by drawing a straight line from 2 to 4 to 6 to 8 and so forth, and a final line drawn from 64 to 2.

Suggested project: Draw parts one and two in different colors.

Figure 1

	3		8		6		1	
1	17	40	57	16	41	32	1	56	(8)
2	21	36	61	12	45	28	5	52	(7)
3	19	38	59	14	43	30	3	54	(6)
4	23	34	63	10	47	26	7	50	(5)
5	18	39	58	15	42	31	2	55	(4)
6	22	35	62	11	46	27	6	51	(3)
7	20	37	60	13	44	29	4	53	(2)
8	24	33	64	9	48	25	8	49	(1)
	5		2		4		7	

3	5	8	2
6	4	1	7

Preliminary Square Using Sequence of Complementary Numbers

4	1	7	6
5	8	2	3

4	23	34	63	10	15	58	39	18	5
(3)	46	27	6	51	54	3	30	43	(6)
1	17	40	57	16	9	64	33	24	8
(2)	44	29	4	53	52	5	28	45	(7)
7	20	37	60	13	12	61	36	21	2
(8)	41	32	1	56	49	8	25	48	(1)
6	22	35	62	11	14	59	38	19	3
(5)	47	26	7	50	55	2	31	42	(4)

Magic Square

Figure 2: This sequence design makes use of the plus-four sequence and was made in four parts:

Part one was made by drawing a straight line from 1 to 5 to 9 to 13 to 17 and so forth, and a final line from 61 to 1.

Part two was made by drawing a straight line from 2 to 6 to 10 to 14 and so forth, and a final line from 62 to 2.

Part three was made by drawing a straight line from 3 to 7 to 11 to 15 to 19 and so forth, and a final line from 63 to 3.

Part four was made by drawing a straight line from 4 to 8 to 12 to 16 and so forth, and a final line from 64 to 4.

Suggested project: Draw four parts in different colors.

Figure 2

						1	
1						1	
2						5	
3						3	
4						7	
5						2	
6						6	
7						4	
8						8	

3	5	8	2
6	4	1	7

4	1	7	6
5	8	2	3

23	34	63	10	15	58	39	18
46	27	6	51	54	3	30	43
17	40	57	16	9	64	33	24
44	29	4	53	52	5	28	45
20	37	60	13	12	61	36	21
41	32	1	56	49	8	25	48
22	35	62	11	14	59	38	19
47	26	7	50	55	2	31	42

Magic Square

Figure 3: This sequence design makes use of the plus-eight sequence.

It was made by drawing a straight line from 1 to 9 to 2 to 10 to 3 to 11 to 4 to 12 and so forth. A final line was drawn from 64 to 1.

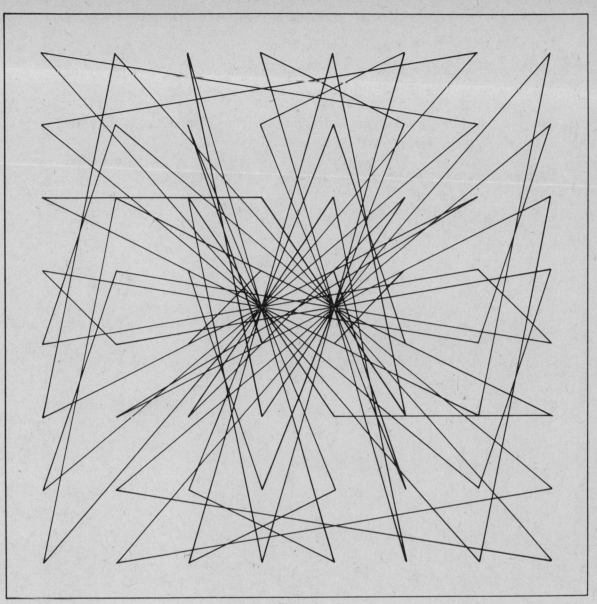

Figure 3

3	5	8	2
6	4	1	7

							1
1							1
2							5
3							3
4							7
5							2
6							6
7							4
8							8

4	1	7	6
5	8	2	3

23	34	63	10	15	58	39	18
46	27	6	51	54	3	30	43
17	40	57	16	9	64	33	24
44	29	4	53	52	5	28	45
20	37	60	13	12	61	36	21
41	32	1	56	49	8	25	48
22	35	62	11	14	59	38	19
47	26	7	50	55	2	31	42

Magic Square

Figure 4: This sequence design makes use of the plus-sixteen sequence and was made in two parts.

Part one was made by drawing a straight line from 1 to 17 to 2 to 18 to 3 to 19 to 4 to 20 and so forth, and a final line from 32 to 1.

Part two was made by drawing a straight line from 33 to 49 to 34 to 50 to 35 to 51 and so forth, and a final line from 64 to 33.

Suggested project: Draw parts one and two in different colors.

Figure 4

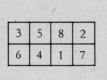

3	5	8	2
6	4	1	7

						1	
1						1	
2						5	
3						3	
4						7	
5						2	
6						6	
7						4	
8						8	

4	1	7	6
5	8	2	3

23	34	63	10	15	58	39	18
46	27	6	51	54	3	30	43
17	40	57	16	9	64	33	24
44	29	4	53	52	5	28	45
20	37	60	13	12	61	36	21
41	32	1	56	49	8	25	48
22	35	62	11	14	59	38	19
47	26	7	50	55	2	31	42

Magic Square

Figure 5: This sequence design makes use of the plus-thirty-two sequence.

It was made by drawing a straight line from 1 to 33 to 2 to 34 to 3 to 35 and so forth, and a final line from 64 to 1.

Figure 5

							1
1							1
2							5
3							3
4							7
5							2
6							6
7							4
8							8

3	5	8	2
6	4	1	7

4	1	7	6
5	8	2	3

23	34	64	10	15	58	39	18
46	27	6	51	54	3	30	43
17	40	57	16	9	64	33	24
44	29	4	53	52	5	28	45
20	37	60	13	12	61	36	21
41	32	1	56	49	8	25	48
22	35	62	11	14	59	38	19
47	26	7	50	55	2	31	42

Magic Square

This is a purposely unfinished sequence design of figure 5 and is shown to emphasize the interesting design possibilities of magic squares.

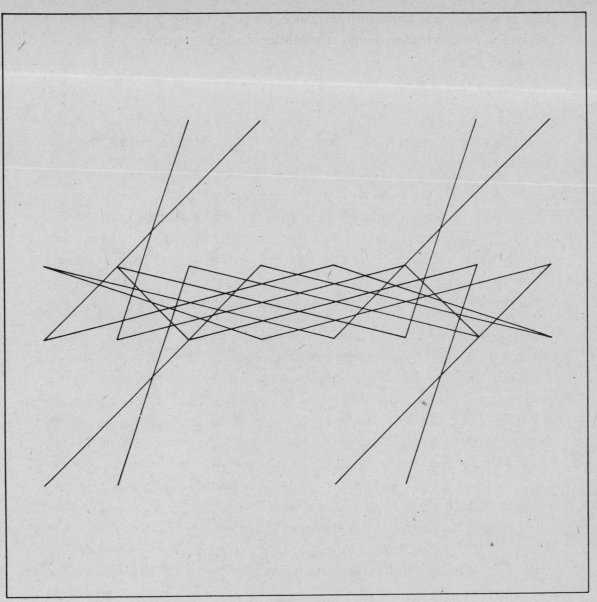

Figure 5A

						1	
1						1	
2						5	
3						3	
4						7	
5						2	
6						6	
7						4	
8						8	

3	5	8	2
6	4	1	7

4	1	7	6
5	8	2	3

23	34	63	10	15	58	39	18
46	27	6	51	54	3	30	43
17	40	57	16	9	64	33	24
44	29	4	53	52	5	28	45
20	37	60	13	12	61	36	21
41	32	1	56	49	8	25	48
22	35	62	11	14	59	38	19
47	26	7	50	55	2	31	42

Magic Square

BENJAMIN FRANKLIN ON MAGIC SQUARES

LETTER TO PETER COLLINSON (C. 1750)

Sir,

According to your request, I now sent you the arithmetical curiosity, of which this is the history.

Being one day in the country, at the house of our common friend, the late learned Mr. Logan, he shewed me a folio French book, filled with magic squares . . . in which he said, the author had discovered great ingenuity and dexterity in the management of numbers; and, though several other foreigners had distinguished themselves in the same way, he did not recollect that any one Englishman had done anything of the kind remarkable.

. . . I then confessed to him, that in my younger days, having once some leisure (which I still think I might have employed more usefully), I had amused myself in making these kind of magic squares, and, at length, had acquired such a knack at it, that I could fill the cells of any magic square, of reasonable size, with a series of numbers as fast as I could write them, disposed in such a manner, as that the sums of every row, horizontal, perpendicular, or diagonal, should be equal; but not being satisfied with these, which I looked on as common and easy things, I had imposed on myself more difficult tasks, and succeeded in making other magic squares, with a variety of properties, and much more curious. He then shewed me several in the same book, of an uncommon and more curious kind; but, as I thought none of them equal to some I remembered to have made, he desired me to let him see them; and accordingly, the next time I visited him, I carried him a square of 8, which I found among my old papers, and which I will now give you, with an account of its properties [see figure 1].

The properties are,

1. That every straight row (horizontal or vertical) of 8 numbers added together, makes 260, and half each row half 260.

2. That the bent row of 8 numbers, ascending and descending diagonally, viz. from 16 ascending to 10, and from 23 descending to 17; and every one of its parallel bent rows of 8 numbers, make 260. Also the bent row from 52, descending to 54, and from 43 ascending to 45; and every one of its parallel bent rows of 8 numbers, make 260. Also the bent row from 45 to 43 descending to the left, and from 23 to 17 descending to the right, and every one of its parallel bent

rows of 8 numbers, make 260. Also the bent row from 52 to 54 descending to the right, and from 10 to 16 descending to the left, and every one of its parallel bent rows of 8 numbers, make 260. Also the parallel bent rows next to the above-mentioned, which are shortened to 3 numbers ascending, and 3 descending, etc., as from 53 to 4 ascending, and from 29 to 44 descending, make with the 2 corner numbers, 260. Also the 2 numbers, 14, 61 ascending, and 36, 19, descending, with the lower 4 numbers situated like them, viz. 50, 1 descending, and 32, 47 ascending, make 260. And, lastly, the 4 corner numbers with the 4 middle numbers make 260. So this magical square seems perfect in its kind.

But these are not all its properties; there are 5 other curious ones, which, at some other time, I will explain to you.

52	61	4	13	20	29	36	45
14	3	62	51	46	35	30	19
53	60	5	12	21	28	37	44
11	6	59	54	43	38	27	22
55	58	7	10	23	26	39	42
9	8	57	56	41	40	25	24
50	63	2	15	18	31	34	47
16	1	64	49	48	33	32	17

Figure 1

Properties of Franklin's 8 X 8 Square

Patterns below indicate individual number locations within the Ben Franklin 8 × 8 square.

Patterns A, B, D, E, F, and G may be moved up and down within the square and still indicate the same totals.

Patterns H, I, J, and K may be moved to left or right within the square and still indicate the same totals.

Pattern C may be moved anywhere within the square and still indicate the same total.

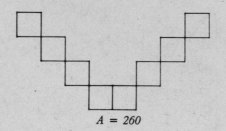

A = 260

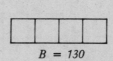

B = 130

C = 130

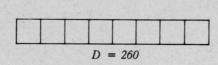

D = 260

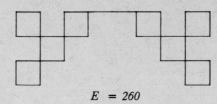

E = 260

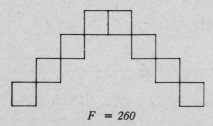

F = 260

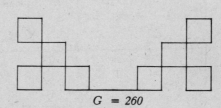

G = 260

$H = 130$

$I = 260$

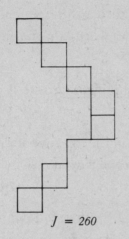

$J = 260$

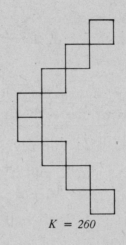

$K = 260$

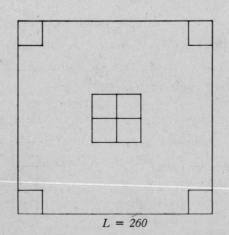

$L = 260$

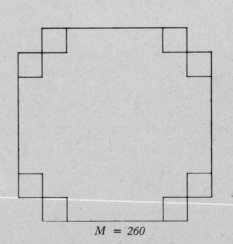

$M = 260$

AN ATTEMPT TO CRACK FRANKLIN'S SECRET 8 X 8 FORMULA

Ben Franklin does not explain the methods he used to construct his two known squares, but it is reasonable to suppose that the method used for his 8 × 8 was based on a more complicated variation of the one shown below.

The simple construction of preliminary square A is shown in figure 1 (see opposite page).

Figure 2, the first step of preliminary square B, shows the initial placement of a 1 through 8 sequence, which, beginning with 1 in the upper left corner, zigzags downward, alternating between columns one and two.

Figure 3 shows the second placement of the 1 through 8 sequence, which, beginning with 1 in the lower left corner, zigzags upward, alternating between columns one and two.

Figure 4 shows the remaining six columns of preliminary square B completed in the same manner as the first two columns.

The alphabetical sequence A through H is placed in the top row of preliminary square B. The same sequence is reversed in the second row and so on, as shown completed in figure 4.

Square C is made up of preliminary squares A and B in the following manner: the upper left corner of square B (figure 4) shows the index number A1. This refers to the A1 position in square A (figure 1), which is the upper left corner box. This box contains the number 1. The number 1 is placed in the upper left corner box of square C.

Continuing with the top row of square B (figure 4), the second box shows the index number B8. This refers to the B8 box of square A, which contains the number 16. 16 is placed in the second box of the top row of square C. The remaining boxes of square C are filled in the same manner.

The completed square C (figure 5) made by the above method of simple alternation is not a true magic square by classic definition, but it does have eight out of the thirteen magical properties (see A, C, D, E, F, G, L, and M) of Franklin's 8 × 8, which shows that we are on the right track in our attempt to discover Franklin's method. It also has an astonishing symmetry, which is revealed by the sequence design below.

Figure 1
Plan of Construction
Preliminary Square A

	1	2	3	4	5	6	7	8	
A	1	2	3	4	5	6	7	8	A
B	9	10	11	12	13	14	15	16	B
C	17	18	19	20	21	22	23	24	C
D	25	26	27	28	29	30	31	32	D
E	33	34	35	36	37	38	39	40	E
F	41	42	43	44	45	46	47	48	F
G	49	50	51	52	53	54	55	56	G
H	57	58	59	60	61	62	63	64	H
	1	2	3	4	5	6	7	8	

Figure 2
First Step of Preliminary Square B

A_1	B	C	D	E	F	G	H
H	G_2	F	E	D	C	B	A
$_3$							
	$_4$						
$_5$							
	$_6$						
$_7$							
	$_8$						

Figure 3
Second Step of Preliminary Square B

A_1	B_8	C	D	E	F	G	H
H_7	G_2	F	E	D	C	B	A
$_3$	$_6$						
$_5$	$_4$						
$_5$	$_4$						
$_3$	$_6$						
$_7$	$_2$						
$_1$	$_8$						

Figure 4
Completed Preliminary Square B

A_1	B_8	C_1	D_8	E_1	F_8	G_1	H_8
H_7	G_2	F_7	E_2	D_7	C_2	B_7	A_2
A_3	B_6	C_3	D_6	E_3	F_6	G_3	H_6
H_5	G_4	F_5	E_4	D_5	C_4	B_5	A_4
A_5	B_4	C_5	D_4	E_5	F_4	G_5	H_4
H_3	G_6	F_3	E_6	D_3	C_6	B_3	A_6
A_7	B_2	C_7	D_2	E_7	F_2	G_7	H_2
H_1	G_8	F_1	E_8	D_1	C_8	B_1	A_8

Figure 5
Square C Made Up from Preliminary Squares A and B

1	16	17	32	33	48	49	64
63	50	47	34	31	18	15	2
3	14	19	30	35	46	51	62
61	52	45	36	29	20	13	4
5	12	21	28	37	44	53	60
59	54	43	38	27	22	11	6
7	10	23	26	39	42	55	58
57	56	41	40	25	24	9	8

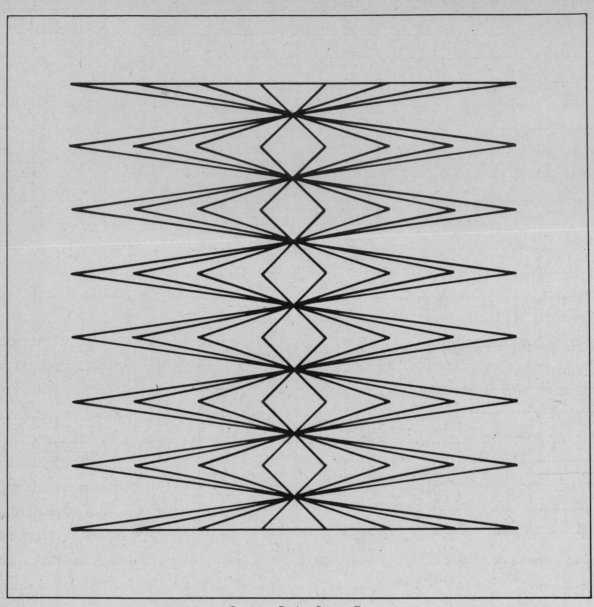

Sequence Design Square C
(See Figure 5)

A BETTER ATTEMPT TO CRACK FRANKLIN'S SECRET 8 X 8 FORMULA

Figure 7, preliminary square A, is the same as previous figure 1.

Figure 8, the first step of preliminary square B, shows the first two columns divided horizontally in half.

In the upper half, the 1 through 4 sequence, beginning with 1 in the upper left corner, zigzags downward, alternating between columns one and two. The 5 through 8 sequence zigzags upward, filling in the remaining unfilled boxes in the upper half.

The numbers in the lower half are the same as those in the upper half except that the paired digits are reversed.

The two columns shown in figure 8 have further been divided into four sections, each containing four numbers, as indicated by the heavy lines.

The upper section is now shifted horizontally across to figure 9, as indicated by the arrow. The other three sections are also shifted to figure 9, but to different positions as indicated by the arrows.

This final arrangement of the numbers in columns one and two is repeated (as shown in figure 10) to complete preliminary square B.

Square C (figure 11) is made by the same method as the previous square C (figure 5).

	1	2	3	4	5	6	7	8	
A	1	2	3	4	5	6	7	8	A
B	9	10	11	12	13	14	15	16	B
C	17	18	19	20	21	22	23	24	C
D	25	26	27	28	29	30	31	32	D
E	33	34	35	36	37	38	39	40	E
F	41	42	43	44	45	46	47	48	F
G	49	50	51	52	53	54	55	56	G
H	57	58	59	60	61	62	63	64	H
	1	2	3	4	5	6	7	8	

Figure 7
Plan of Construction
Preliminary Square A (Same As Previous Figure 1)

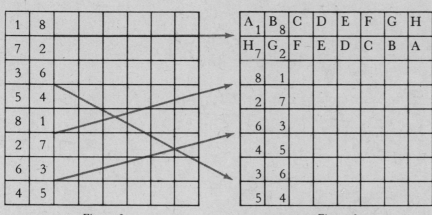

Figure 8
First Step of Preliminary Square B

Figure 9
Second Step of Preliminary Square B

Figure 10
Completed Preliminary Square B

1	16	17	32	33	48	49	64
63	50	47	34	31	18	15	2
8	9	24	25	40	41	56	57
58	55	42	39	26	23	10	7
6	11	22	27	38	43	54	59
60	53	44	37	28	21	12	5
3	14	19	30	35	46	51	62
61	52	45	36	29	20	13	4

Figure 11
Square C

It took a little systematic juggling and shifting of the numbers, but our new square C (figure 11) has all the properties of Franklin's 8 × 8 except one (see B).

Note similarities of sequence design below with the sequence design of Franklin's 8 × 8. (Design has been placed upside down because it looks better that way.)

Square C
Figure 11 in Upside-Down Position

THE CRACKING OF FRANKLIN'S
SECRET 8 X 8 FORMULA

We'll never know for sure how Dr. Franklin put together his fantastic 8 × 8. All we can say is "Bravo, Ben!" As he states in his letter to Peter Collinson, *he had the knack of arranging the numbers in magic order as fast as he could write them down*, and there is no reason to disbelieve him. (Or is there?) He undoubtedly used a method far superior to the rather laborious one presented here, but we are at least able to show one possible approach.

Square C (figure 11) is used as the key to Franklin's 8 × 8, as may be seen below.

Square D (figure 12) is made by reversing the order of the numbers in all eight columns of square C.

The first step of Square E is made by interchanging the positions of columns one and two of square D.

The second step of square E is made by interchanging the positions of columns three and eight of square D.

The third step of square E is made by interchanging the positions of columns four and seven of square D.

The fourth and final step of square E is made by interchanging the positions of columns five and six of square D, which completes square E, which is Franklin's 8 × 8.

1	16	17	32	33	48	49	64
63	50	47	34	31	18	15	2
8	9	24	25	40	41	56	57
58	55	42	39	26	23	10	7
6	11	22	27	38	43	54	59
60	53	44	37	28	21	12	5
3	14	19	30	35	46	51	62
61	52	45	36	29	20	13	4

Square C (Figure 11)

61	52	45	36	29	20	13	4
3	14	19	30	35	46	51	62
60	53	44	37	28	21	12	5
6	11	22	27	38	43	54	59
58	55	42	39	26	23	10	7
8	9	24	25	40	41	56	57
63	50	47	34	31	18	15	2
1	16	17	32	33	48	49	64

Square D (Figure 12)

52	61						
14	3						
53	60						
11	6						
55	58						
9	8						
50	63						
16	1						

Square E—First Step

52	61	4					45
14	3	62					19
53	60	5					44
11	6	59					22
55	58	7					42
9	8	57					24
50	63	2					47
16	1	64					17

Square E—Second Step

52	61	4	13			36	45
14	3	62	51			30	19
53	60	5	12			37	44
11	6	59	54			27	22
55	58	7	10			39	42
9	8	57	56			25	24
50	63	2	15			34	47
16	1	64	49			32	17

Square E—Third Step

52	61	4	13	20	29	36	45
14	3	62	51	46	35	30	19
53	60	5	12	21	28	37	44
11	6	59	54	43	38	27	22
55	58	7	10	23	26	39	42
9	8	57	56	41	40	25	24
50	63	2	15	18	31	34	47
16	1	64	49	48	33	32	17

Square E—Fourth Step:
The Franklin 8 × 8

SEQUENCE DESIGN OF FRANKLIN'S 8 X 8

A CONTINUATION OF FRANKLIN'S LETTER TO COLLINSON

... I went home, and made, that evening, the following magical square of 16, which, besides having all the properties of the foregoing square of eight, i.e. it would make the 2056 in all the same rows and diagonals, had this added, that a four square hole being cut in a piece of paper of such a size as to take in and shew through it, just 16 of the little squares, when laid on the greater square, the sum of the 16 numbers so appearing through the hole, wherever it was placed on the greater square, should likewise make 2056. This I sent to our friend the next morning, who, after some days, sent it back in a letter with these words: "I return to thee thy astonishing or most stupendous piece of the magical square, in which"—but the compliment is too extravagant, and therefore, for his sake, as well as my own, I ought not to repeat it. Nor is it necessary; for I make no question but you will readily allow this square of 16 to be the most magically magical of any magic square ever made by any magician [see figure 2].

B. Franklin

200	217	232	249	8	25	40	57	72	89	104	121	136	153	168	185
58	39	26	7	250	231	218	199	186	167	154	135	122	103	90	71
198	219	230	251	6	27	38	59	70	91	102	123	134	155	166	187
60	37	28	5	252	229	220	197	188	165	156	133	124	101	92	69
201	216	233	248	9	24	41	56	73	88	105	120	137	152	169	184
55	42	23	10	247	234	215	202	183	170	151	138	119	106	87	74
203	214	235	246	11	22	43	54	75	86	107	118	139	150	171	182
53	44	21	12	245	236	213	204	181	172	149	140	117	108	85	76
205	212	237	244	13	20	45	52	77	84	109	116	141	148	173	180
51	46	19	14	243	238	211	206	179	174	147	142	115	110	83	78
207	210	239	242	15	18	47	50	79	82	111	114	143	146	175	178
49	48	17	16	241	240	209	208	177	176	145	144	113	112	81	80
196	221	228	253	4	29	36	61	68	93	100	125	132	157	164	189
62	35	30	3	254	227	222	195	190	163	158	131	126	99	94	67
194	223	226	251	2	31	34	63	66	95	98	127	130	159	162	191
64	33	32	1	256	225	224	193	192	161	160	129	128	97	96	65

Figure 2

P.S. If you really want to take a trip, make your own sequence design of the fabulous magic square.

Ben Franklin was undoubtedly well aware of the ground rules governing magic squares—that is, that each of the rows, columns, and corner-to-corner diagonals must add up to the same total. It is typical of the genius of Franklin that he ignored the classic ground rules, cut out on his own, and came up with some fantastic results, as may be seen by an analysis of his two squares. The corner-to-corner diagonals are not magic, but this didn't seem to bug freak Franklin. The new features he added more than made up for this deficiency, and he was justly qualified to make his proud boast that his 16 × 16 is "the most magically magical of any magic square ever made by any magician."

LETTER TO THE AUTHOR FROM HERB THE FALCONER (KAHN)

Dear Jim:

Thank you for sending me your stuff on Ben Franklin and his 8 × 8 and 16 × 16 magic squares. These squares are indeed marvelous and he had every reason to be justly proud of them. But why did he have to write that boastful letter to Peter Collinson?

Ben has always been one of my idols, and now he has gone and tarnished his image and revealed himself to me as being merely human. In telling Collinson that he could make his magic squares as fast as he could write down the numbers, he was pushing reasonable exaggeration beyond the limit. I just can't buy this, and I say nobody but *nobody* can perform this feat. He even claimed he could do this with squares "of any reasonable size." *No way!* I'd like to see him or anybody else tackle a 6 × 6 or a 10 × 10 in this fashion. I should have known better than to idolize Ben. After all, Ben was a distinguished career diplomat, and what career diplomat ever became distinguished who wasn't gifted in the arts of deception?

Another thing which may account for Ben's lapse of rectitude is that he was vain—he admits this in his autobiography. He tells of a little incident which happened during his early life in Philadelphia. He was walking down the street one day and encountered one of the Quaker elders, who among other things told him: "We think thou art guilty of the sin of pride." This hurt Ben deeply, and when he got home he made a list of all his shortcomings: pride, vanity, gluttony, sloth, avarice, lust, etc., etc.—and resolved to conquer them quickly. Each night at bedtime, he crossed off those that he had licked that day. At the end of the month, he had vanquished them all and found that he was very *proud* of his accomplishment. And there he was, back at square one.

By the way, it took me all afternoon, but I made a sequence design of Ben's 16 × 16. The way this thing zigzags around to weave its pattern blows my mind. It's the most dazzling yet.

I'm still trying to crack the 6 × 6, but no luck yet. But *don't tell me,* I want to solve this one all by myself.

Keep in touch,

Herb

P.S. Hey! I discovered a new and easier way to find the constant of a magic square of any size. I think it's the best way yet. Get a load of this!

A. Lay out all the numbers of a 3 × 3 in a straight line. Circle the three middle numbers and add them up. The total is 15 (constant).
B. Do the same with a 4 × 4 and circle the four middle numbers. They total 34 (constant).
C. 5 × 5: Circle the five middle numbers and add them up.
D. 6 × 6: Circle the six middle numbers and add them up.

3 × 3:   1 2 3 4 5 6 7 8 9   4 + 5 + 6 = 15 (constant)
4 × 4:   1 2 3 4 5 6 7 8 9 10 11 12 13 14 15 16   7 + 8 + 9 + 10 = 34 (constant)

Etc., Etc. Okay?

H

METHOD FOR CONSTRUCTING
10 X 10 MAGIC SQUARE

1. Figure 1 shows the sequence of 1 through 100 arranged in its natural order.
2. In figure 2, the numbers within the central 4 × 4 square of figure 1 have been rearranged by interchanging the positions of complementary pairs of numbers. This is the final arrangement of these particular numbers, as may be seen in the completed 10 × 10 magic square (figure 8, page 210).
3. In figure 3, the numbers in the central 4 × 4 have been excluded and the order of the remaining numbers in the four central rows and columns of figure 1 has systematically switched as shown.
4. Figure 4 shows the alteration of figure 1 brought about by the changes made in figures 2 and 3.
5. Figure 5 shows what might be called a symmetrical switch pattern, the purpose of which will be explained later.
6. Figure 6 shows the same switch pattern turned 90 degrees counterclockwise.
7. Figure 7 combines figures 5 and 6 to show the completed switch pattern.
8. The switch pattern in figure 7 is used to rearrange the numbers in figure 4 in the following way to complete the 10 × 10 magic square shown in figure 8:
   Note that the bottom row of figure 7 contains two horizontal dashes. These two horizontal dashes refer to the two corresponding boxes in the bottom row of figure 4 and indicate that the numbers (5 and 6) must be interchanged (switched), as shown in figure 8.
9. Note that the top row of figure 7 also contains two horizontal dashes, which again refer to the two corresponding boxes in the top row of figure 4 and indicate that the numbers (2 and 9) must be interchanged (see figure 8).
10. This process is continued with each pair of horizontal dashes in the other rows (five pairs in all).
11. The vertical dashes in figure 7 are used in the same way to indicate vertical interchanges of numbers in figure 4 (see figure 8).
12. Figure 8 shows the completed magic square after all of the above interchanges have been made.
13. Figure 9 shows another switch pattern similar to figure 5, which may be used as the basis for the construction of a different 10 × 10 magic square.

Suggested project: Try to devise other switch patterns that may be used to construct additional 10 × 10 magic squares.

Figure 1:

1	2	3	4	5	6	7	8	9	10
11	12	13	14	15	16	17	18	19	20
21	22	23	24	25	26	27	28	29	30
31	32	33	34	35	36	37	38	39	40
41	42	43	44	45	46	47	48	49	50
51	52	53	54	55	56	57	58	59	60
61	62	63	64	65	66	67	68	69	70
71	72	73	74	75	76	77	78	79	80
81	82	83	84	85	86	87	88	89	90
91	92	93	94	95	96	97	98	99	100

Figure 1

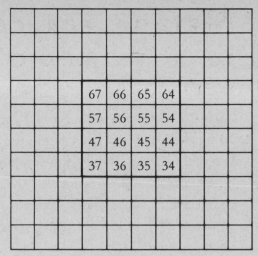

Figure 2

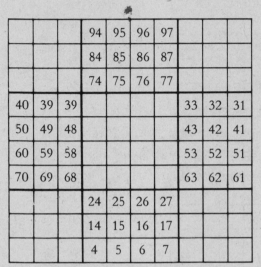

Figure 3

Figure 4:

1	2	3	94	95	96	97	8	9	10
11	12	13	84	85	86	87	18	19	20
21	22	23	74	75	76	77	28	29	30
40	39	38	67	66	65	64	33	32	31
50	49	48	57	56	55	54	43	42	41
60	59	58	47	46	45	44	53	52	51
70	69	68	37	36	35	34	63	62	61
71	72	73	24	25	26	27	78	79	80
81	82	83	14	15	16	17	88	89	90
91	92	93	4	5	6	7	98	99	100

Figure 4

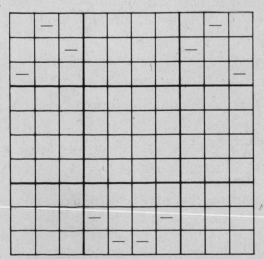

Figure 5

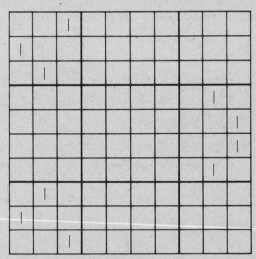

Figure 6

Figure 7

1	9	93	94	95	96	97	8	2	10
81	12	18	84	85	86	87	13	19	20
30	72	23	74	75	76	77	28	29	21
40	39	38	67	66	65	64	33	62	31
50	49	48	57	56	55	54	43	42	51
60	59	58	47	46	45	44	53	52	41
70	69	68	37	36	35	34	63	32	61
71	22	73	24	25	26	27	78	79	80
11	82	83	17	15	16	14	88	89	90
91	92	3	4	5	6	7	98	99	100

Figure 8

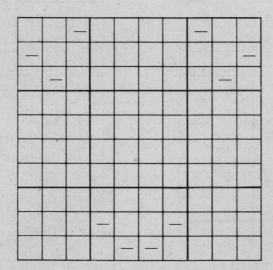

Figure 9

METHOD FOR CONSTRUCTING
12 X 12 COMPOSITE MAGIC SQUARES

The 12 × 12 composite magic square as shown in figure 4 may be constructed in the following manner:

1. Construct 3 × 3 magic square (see figure 1).
2. Construct 4 × 4 magic square (see figure 2).
3. Divide 12 × 12 diagram into sixteen 3 × 3 squares (see figure 3).
4. Number the sixteen 3 × 3 squares (see large numbers in figure 3) using the same arrangement of numbers as was used in the 4 × 4 magic square (figure 2).
5. Divide the 12 × 12 series (1 through 144) into groups of nine numbers: 1–9, 10–18, 19–27, etc.
6. Using the same arrangement of numbers as was used in the 3 × 3 magic square (figure 1), place the first group of nine numbers (1–9) in the lower right corner 3 × 3 division of the 12 × 12 (see figure 3). This position corresponds to the lower right corner box of the 4 × 4 magic square (figure 2) which contains the number 1 indicating the first position.
7. Place the second group of nine numbers (10–18) in the 12 × 12 square to occupy the position corresponding to the location of the number 2 in the 4 × 4 magic square. Note that the second group of nine numbers is arranged in the same order as the 3 × 3 magic square.
8. The third group of nine (19–27) is placed within the 12 × 12 in the position indicated by the number 3 in the 4 × 4 magic square (figure 2).
9. This procedure is continued until the 12 × 12 magic square is completed, as shown in figure 4.

8	1	6
3	5	7
4	9	2

Figure 1
3 × 3 Magic Square

16	2	3	13
5	11	10	8
9	7	6	12
4	14	15	1

Figure 2
4 × 4 Magic Square

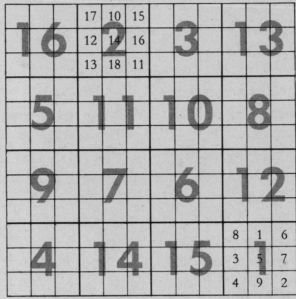

Figure 3

143	136	141	17	10	15	26	19	24	116	109	114
138	140	142	12	14	16	21	23	25	111	113	115
139	144	137	13	18	11	22	27	20	112	117	110
44	37	42	98	91	96	89	82	87	71	64	69
39	41	43	93	95	97	84	86	88	66	68	70
40	45	38	94	99	92	85	90	83	67	72	65
80	73	78	62	55	60	53	46	51	107	100	105
75	77	79	57	59	61	48	50	52	102	104	106
76	81	74	58	63	56	49	54	47	103	108	101
35	28	33	125	118	123	134	127	132	8	1	6
30	32	34	120	122	124	129	131	133	3	5	7
31	36	29	121	126	119	130	135	128	4	9	2

Figure 4
(Constant 870)
12 × 12 Magic Square Made with Sixteen 3 × 3's

A DIFFERENT METHOD FOR CONSTRUCTING
12 × 12 COMPOSITE MAGIC SQUARES

Figures 5, 6, and 7 show a different construction pattern for 12 × 12 composite magic squares. The general idea is the same as the previous method, but the 12 × 12 square (figure 7) is divided into nine 4 × 4 squares which are numbered in big letters, corresponding to the 3 × 3 magic square (figure 5).

The 12 × 12 magic square may be completed by following the procedure begun with figures 5, 6, and 7.

8	1	6
3	5	7
4	9	2

Figure 5

16	2	11	5
3	13	8	10
6	12	1	15
9	7	14	4

Figure 6

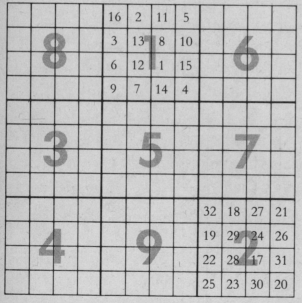

Figure 7
12 × 12 Made with Nine 4 × 4's

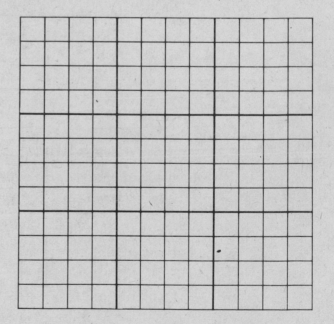

WHAT GOOD IS THE BINARY SYSTEM?
WHAT GOOD ARE MAGIC SQUARES?

In the binary system of notation, zeros and ones only are used to express any number. For example, the number 25 is written as 11001 in binary. Like magic squares, the binary system has existed for centuries, and up until about the late thirties and early forties was considered useless from any practical standpoint— at best a musty mathematical recreation.

Then suddenly a practical application was discovered—an application so simple it could have been conceived by a ten-year-old, yet one of such overwhelming magnitude and importance that it changed our world. The old binary system was dusted off and applied to the electronic computer and made the space age possible.

It is true that the binary system has built-in disadvantages if we try to use it for everyday home accounting purposes. We wouldn't want to have to write out a row of seven zeros and ones to express the number 25. But binary is ideal for the computer. The computer doesn't care how long a row of zeros and ones it takes to express a given number, but the principal advantage to the electronic computer is that the zeros and ones may be used to indicate off or on, circuit open or closed, yes and no, true and false, and so forth.

We have answered the question "What good is the binary system?" It lay around in antique manuscripts half forgotten for centuries and then wham! A brilliant application was discovered.

"What good are magic squares?" We're not sure yet, but we certainly cannot discount the possibility that there may be some budding genius among us who is going to come up with the "Eureka!"

THE STRUCTURE OF THE BINARY SYSTEM

1. Note that column A is a series of alternating zeros and ones.
2. Note that column B begins *two* spaces lower than column A and is an alternating series of *two* ones and *two* zeros.
3. Note that column C begins *four* spaces lower than column A and is an alternating series of *four* ones and *four* zeros.
4. Note that column D begins *eight* spaces lower than column A and is an alternating series of *eight* ones and *eight* zeros.
5. Note that column E begins *sixteen* spaces lower than column A and is an alternating series of *sixteen* ones and *sixteen* zeros.
6. Column F will begin *thirty-two* spaces lower than column A and (if continued) would become an alternating series of *thirty-two* ones and *thirty-two* zeros.
7. It may be seen that by continuing the above pattern of progression, any number can be expressed in binary notation.

		Binary					Decimal
Column A →					0	=	0
					1	=	1
Column B →				1	0	=	2
				1	1	=	3
Column C →			1	0	0	=	4
			1	0	1	=	5
			1	1	0	=	6
			1	1	1	=	7
Column D →		1	0	0	0	=	8
		1	0	0	1	=	9
		1	0	1	0	=	10
		1	0	1	1	=	11
		1	1	0	0	=	12
		1	1	0	1	=	13
		1	1	1	0	=	14
		1	1	1	1	=	15
Column E →	1	0	0	0	0	=	16
	1	0	0	0	1	=	17
	1	0	0	1	0	=	18
	1	0	0	1	1	=	19
	1	0	1	0	0	=	20
	1	0	1	0	1	=	21
	1	0	1	1	0	=	22
	1	0	1	1	1	=	23
	1	1	0	0	0	=	24
	1	1	0	0	1	=	25
	1	1	0	1	0	=	26
	1	1	0	1	1	=	27
	1	1	1	0	0	=	28
	1	1	1	0	1	=	29
	1	1	1	1	0	=	30
	1	1	1	1	1	=	31

Column F → 1 0 0 0 0 0 = 32

Who's afraid of the 6 × 6?

If you have progressed this far in the book and have absorbed the many methods of constructing magic squares of various sizes, you might reasonably assume that you would have no trouble in constructing a magic square of any size. Also you may have noticed the omission of any method for constructing a 6 × 6. This was deliberately left out because we didn't want to deprive you of the joy of figuring out a method of your own.

Offhand the 6 × 6 may not seem like much of a challenge, but we'll tell you in advance that when you tackle this, you'll be opening up a can of worms. If you can do it, you are a magic-square freak first class (M.S.F.F.C.). Here are a couple of sample cans (see figures 1 and 2 below). How were they made?

Good luck!

34	6	5	32	1	33
9	25	11	8	30	28
15	18	20	23	19	16
22	24	14	17	13	21
27	7	26	29	12	10
4	31	35	2	36	3

Figure 1

36	4	2	35	3	31
7	27	8	11	28	30
13	16	23	20	21	18
24	22	17	14	15	19
25	9	29	26	10	12
6	33	32	5	34	1

Figure 2

PECULIARITIES OF THE NUMBER 9

Take any number—let's say 125436.

Add up the digits: $1 + 2 + 5 + 4 + 3 + 6 = 21$.

Subtract this total (21) from the original number.

$$\begin{array}{r} 125436 \\ -\ 21 \\ \hline 125415 \end{array}$$

Add up these digits: $1 + 2 + 5 + 4 + 1 + 5 = 18$.

Add up these digits (18): $1 + 8 = 9$.

You will always end up with 9 no matter what number you begin with.

Take any number—let's say 70642.

Reverse these digits. We get 24607.

Subtract the smaller from the larger $= 46035$.

Add up the digits: $4 + 6 + 0 + 3 + 5 = 18$.

Add up these digits (18): $1 + 8 = 9$.

You will always end up with 9 no matter what number you begin with.

$$
\begin{array}{r}
0 \times 9 + 1 = 1\ 0 \\
1 \times 9 + 2 = 1\ 1 \\
1\ 2 \times 9 + 3 = 1\ 1\ 1 \\
1\ 2\ 3 \times 9 + 4 = 1\ 1\ 1\ 1 \\
1\ 2\ 3\ 4 \times 9 + 5 = 1\ 1\ 1\ 1\ 1 \\
1\ 2\ 3\ 4\ 5 \times 9 + 6 = 1\ 1\ 1\ 1\ 1\ 1 \\
1\ 2\ 3\ 4\ 5\ 6 \times 9 + 7 = 1\ 1\ 1\ 1\ 1\ 1\ 1 \\
1\ 2\ 3\ 4\ 5\ 6\ 7 \times 9 + 8 = 1\ 1\ 1\ 1\ 1\ 1\ 1\ 1 \\
1\ 2\ 3\ 4\ 5\ 6\ 7\ 8 \times 9 + 9 = 1\ 1\ 1\ 1\ 1\ 1\ 1\ 1\ 1 \\
1\ 2\ 3\ 4\ 5\ 6\ 7\ 8\ 9 \times 9 + 10 = 1\ 1\ 1\ 1\ 1\ 1\ 1\ 1\ 1\ 1
\end{array}
$$

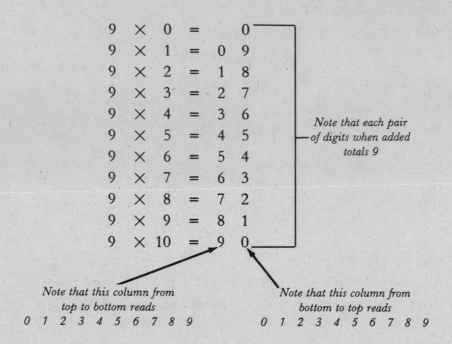

$$9 \times 0 = 0$$
$$9 \times 1 = 0\ 9$$
$$9 \times 2 = 1\ 8$$
$$9 \times 3 = 2\ 7$$
$$9 \times 4 = 3\ 6$$
$$9 \times 5 = 4\ 5$$
$$9 \times 6 = 5\ 4$$
$$9 \times 7 = 6\ 3$$
$$9 \times 8 = 7\ 2$$
$$9 \times 9 = 8\ 1$$
$$9 \times 10 = 9\ 0$$

Note that each pair of digits when added totals 9

Note that this column from top to bottom reads
0 1 2 3 4 5 6 7 8 9

Note that this column from bottom to top reads
0 1 2 3 4 5 6 7 8 9

Multiply any number by 9—let's say 604.

$604 \times 9 = 5436$.

Add up these digits (5436): $5 + 4 + 3 + 6 = 18$.

Add up these digits (18): $1 + 8 = 9$.

You'll always end up with 9.

Take the number 12345679. (Note that the 8 is left out of the 1 through 9 series. Don't ask me why.)

Multiply this number by any multiple of 9 from 9 through 81 (18, 27, 36, 45, 54, etc.) and you'll get a row of identical digits.

Let's try 45 (9 × 5).

$12345679 \times 45 = 555555555$.

9×4 (36) will give you all 4's.

9×7 (63) will give you all 7's.

Try this on your pocket calculator.

Take any three-digit number, provided the first and last digits are not alike—
let's say 764.

Reverse the digits. We get 467.

Subtract the smaller from the larger:

$$
\begin{array}{r}
764 \\
-\ 467 \\
\hline
297
\end{array}
$$

The middle digit will *always* be 9.

The two outside numbers will *always* add up to 9.

$$
\begin{array}{rrrrr}
694 & 825 & 801 & 521 & 761 \\
-\ 496 & -\ 528 & -\ 108 & -\ 125 & -\ 167 \\
\hline
198 & 297 & 693 & 396 & 594
\end{array}
$$

9 million divided by 9 = 1,000,000

8 million divided by 9 = 888,888

7 million divided by 9 = 777,777

6 million divided by 9 = 666,666

etc.

PECULIARITIES OF THE NUMBER 37

37 is an odd number in more ways than one.

Multiply it by any multiple of 3 (up to 27) and here's what happens:

37	37	37	37	37	37	37	37	37
× 3	× 6	× 9	× 12	× 15	× 18	× 21	× 24	× 27
111	222	333	444	555	666	777	888	999

It also takes on other peculiarities when multiplied by higher multiples of 3.

$$37 \times 30 = 1110$$
$$37 \times 33 = 1221$$
$$37 \times 36 = 1332$$
$$37 \times 39 = 1443$$
$$37 \times 42 = 1554$$
$$37 \times 45 = 1665$$
$$37 \times 48 = 1776$$
$$37 \times 51 = 1887$$
$$37 \times 54 = 1998$$
$$37 \times 57 = 2109$$
$$37 \times 60 = 2220$$
$$37 \times 63 = 2331$$
$$37 \times 66 = 2442$$
$$37 \times 69 = 2553$$

THE PECULIARITIES OF NUMBERS

$$1 \times 8 + 1 = 9$$
$$12 \times 8 + 2 = 98$$
$$123 \times 8 + 3 = 987$$
$$1234 \times 8 + 4 = 9876$$
$$12345 \times 8 + 5 = 98765$$
$$123456 \times 8 + 6 = 987654$$
$$1234567 \times 8 + 7 = 9876543$$
$$12345678 \times 8 + 8 = 98765432$$
$$123456789 \times 8 + 9 = 987654321$$

All numbers consisting of six equal figures are divisible by 7.

$$111111 \div 7 = 15873$$
$$222222 \div 7 = 31746$$
$$333333 \div 7 = 47619$$
$$444444 \div 7 = 63492$$
$$555555 \div 7 = 79365$$
$$666666 \div 7 = 95238$$
$$777777 \div 7 = 111111$$
$$888888 \div 7 = 126984$$
$$999999 \div 7 = 142857$$

These same numbers are also divisible by 3 and 11.

BIBLIOGRAPHY

Andrews, W. S. *Magic Squares and Cubes*. New York: Dover Publications, 1960 (reprint of 1917 edition with additions).

Andrews, W. S. *Magic Squares and Cubes*. Chicago: Open Court Publishing Company, 1908, 1917.

Austin, Stuart M. and Howard, Ian. "Magic Squares and Cubes." *Creative Computing*, May 1980.

Benson, William H., and Jacoby, Oswald. *New Recreations with Magic Squares*. New York: Dover Publications, 1976.

Encyclopaedia Britannica, "Magic Squares." 1910, 1911.

Gardner, Martin. *The Second Scientific Book of Mathematical Puzzles and Diversions*. New York: Simon & Schuster, 1966.

Kraitchik, Maurice. *Mathematical Recreations*. New York: Dover Publications, 1953.

GLOSSARY

ASSOCIATIVE MAGIC SQUARES: In associative magic squares all complementary pairs of numbers are diametrically opposed and equidistant from the center.

BROKEN DIAGONALS: The five boxes marked A constitute a full or corner-to-corner diagonal. The five boxes marked B constitute a broken diagonal. The five boxes marked C, D, and E respectively also constitute broken diagonals.

A	B	c	d	e
e	A	B	c	d
d	e	A	B	c
c	d	e	A	B
B	c	d	e	A

COMPLEMENTARY NUMBERS: Pairs of numbers equidistant from the center of a natural series of numbers (see pages 23 and 223).

CONSTANT: See page 14.

EVEN-ORDER MAGIC SQUARES: 4×4, 6×6, 8×8, 10×10, and so forth.

FAMILY DESIGNS: A method of graphic classification of magic squares.

NATURAL ORDER, NATURAL SERIES, NATURAL SEQUENCE: 1, 2, 3, 4, 5, 6, 7, 8, 9, and so forth.

ODD-ORDER MAGIC SQUARES: 3×3, 5×5, 7×7, 9×9, 11×11, and so forth.

PANDIAGONAL MAGIC SQUARES: Pandiagonal magic squares are associated magic squares with one added feature: the numbers in the boxes of the broken diagonals add up to the constant of that square.

RELATIVE ORDER: See page 47.

R.R'S (REFLECTIONS-ROTATIONS), DISGUISED MAGIC SQUARES: See pages 32–34.

SEQUENCE DESIGNS: See page 32.